Technology Operating Models for Cloud and Edge

Create your purpose-built distributed operating model for public, hybrid, multicloud, and edge

Ahilan Ponnusamy

Andreas Spanner

BIRMINGHAM—MUMBAI

Technology Operating Models for Cloud and Edge

Group Product Manager: Preet Ahuja
Publishing Product Manager: Surbhi Suman
Senior Editor: Athikho Sapuni Rishana
Technical Editor: Arjun Varma
Copy Editor: Safis Editing
Project Coordinator: Ashwin Dinesh Kharwa
Proofreader: Safis Editing
Indexer: Subalakshmi Govindhan
Production Designer: Shyam Sundar Korumilli
Marketing Coordinator: Rohan Dobhal

First published: July 2023

Production reference: 1130723

Published by Packt Publishing Ltd.
Grosvenor House
11 St Paul's Square
Birmingham
B3 1R

ISBN 978-1-83763-139-1

www.packtpub.com

To my wife, Kalpana, and kids, Arjun and Krishna, for giving me the time and space to work on this book.

– Ahilan Ponnusamy

To my family and friends near and far, small and tall, learning fast and slow, supporting me with my small, medium, big, and crazy ideas. You always have a place in my heart And a special thanks also to all the people who provide continuous inspiration through their technology understanding, pure talent or skills, positive attitude, support and feedback.

– Andreas Spanner

Forewords

This book is very timely. How we operate going forward and how we organize ourselves as IT organizations is a question on many people's minds. The advent of the cloud made many things easier but, in reality, it created even more variability in the technology stacks that organizations deal with. The toolkit of IT professionals keeps getting bigger, with SRE and platform engineering being notable recent additions and new ideas coming up nearly every week.

What better than to have a book that helps put structure to the problem and demystify some of the buzzwords to cut it back down to what really matters? This book will help IT professionals find approaches to structure their organization, and it does so pragmatically and based on real-life experiences. Real life is messy and every organization will need to make decisions in their context. This book will not give a one-size-fits-all answer; it will help you make your own decisions.

I have spent many hours discussing with Andreas how organizations should react to the challenges of modern technologies being adopted in existing organizations and I am really glad to hold in my hands his best thoughts. With this book, he and his co-author, Ahilan, share so many years of experience with you that I am sure you will find many ideas here that will help you and your organization. It does not just cover the organizational model but also the interaction with technology – it is, after all, a symbiotic relationship when done well.

Let this book be one of the guides you use on your road to improvement. It will surely be one of mine and will take a place on my bookshelf next to other excellent books that make up the canon of modern IT thinking.

Mirco Hering,

Global Transformation Lead at Accenture

In the ever-evolving landscape of technology, businesses now find themselves at a pivotal moment. The exponential growth of data generated in real time in an increasingly distributed manner, and the sudden leap forward in the domain of **Artificial Intelligence** (**AI**), with foundation models making massive scaling possible and allowing enterprises to accelerate the integration of advanced AI capabilities into their operating model, represent both a challenge and an opportunity for businesses, which are often already trying to cope with existing digital transformation and cloud technology adoption challenges. It is in this context that *Technology Operating Models for Cloud and Edge* emerges as a guiding light, providing insights and strategies for organizations seeking to navigate this transformative era and build a strong technical foundation based on enterprise open source technologies for rapid adoption of change.

In this book, Andreas and Ahilan explore the complexities of defining a cloud operating model, the nuances of enterprise technology landscapes, and the limitations of traditional approaches and models such as Bimodal IT. You are empowered to forge a new path forward, leveraging a practical roadmap as well as generic, reusable patterns for building consistent and adaptive operating models across cloud and edge environments. Having had the privilege to closely collaborate with both authors in the context of our work advising organizations across a wide range of sectors, I can definitely see the wealth of their collective experience distilled into a set of actionable principles that can support technology strategists and cloud practitioners alike in navigating the technological complexities inherent to defining and delivering cloud operating models, ultimately harnessing the power of distributed architectures and leveraging their inherent advantages to optimize performance, scalability, and resilience. I am looking forward to embedding these principles on the ground in our future transformative engagements.

Vincent Caldeira,

Chief Technology Officer, APAC at Red Hat

Any enterprise can be decomposed into four components: (a) business model, (b) operating model, (c) technology model, and (d) culture. A digital transformation journey, when done properly, attempts to transform all four components. While each of these components is critical, given that we live in the digital era, the transformation of the technology operating model is absolutely imperative if you want to achieve exponential business outcomes.

The cloud had a tectonic impact on technology operating models. It shifted the way technology was architected, designed, and maintained. Cloud adoption has been wide-ranging, and some might even claim that the impact of the cloud qualifies as being planetary scale. However, just doing a lift and shift to the cloud ticks the box but does not help transform, and exponential outcomes will continue to remain elusive. And while enterprises have struggled to demonstrate value from their cloud initiatives, edge computing and architectures have been introduced into the mix, driven by real business use cases and AI-based aspirations.

Ahilan and Andreas provide an excellent framework and approach for enterprises to build their own distributed technology operating models. They sweep up and provide clarity on various types of cloud architectures, as well as encompass edge computing into their approach. I particularly enjoyed the value addition they provide by describing, in a fair bit of detail, open practices from the open practices library, which can be used to build this operating model from within. This collaborative approach will ensure intrinsic buy-in from key stakeholders toward the distributed technology model and go a long way in ensuring its timely adoption. Besides baking their decades of practical experience into the framework, the book also uses a very detailed, anonymized use case as a simulation, to bring out the methods as well as the learning in a real-life setting. While the framework created by the authors comprising streams, dimensions, and items is applied to the technology operating model, it is quite easily extensible to other components of an enterprise's transformation journey, such as the business model and operating culture.

Digital transformation is easier said and read than done. I recommend you pick up *Technology Operating Models for Cloud and Edge* on your next visit or click to the bookstore, consume its content diligently, then apply the learning to real-life scenarios, and significantly enhance the probability of successfully transforming your own technology operating model.

Neetan Chopra,

Chief Digital and Information Officer at IndiGo

Contributors

About the authors

Ahilan Ponnusamy is a GTM specialist for the application platform at Red Hat based in Singapore. He enjoys working with customers to deliver accelerated business outcomes on hybrid cloud architectures and cloud-native application development and delivery practices. Ahilan completed his master's in computer applications from MKU, India in 1999. His work history includes Philips CE in Eindhoven Netherlands, BEA technologies as a member of customer-centric engineering and support in India and USA, pre-sales tech lead for the cloud platform team at Oracle USA, principal platform engineer at Pivotal(VMware), and global architect at Dell Technologies, Singapore.

Originally from Madurai, Tamil Nadu, India, Ahilan currently resides in Singapore with his wife and two boys.

Andreas Spanner is currently working as a chief architect within the CTO organization at Red Hat. Prior to his role as the chief architect for Australia and New Zealand, Andreas worked across the globe in many different industries ranging from automotive, manufacturing, and supply chain logistics to telco, FSI, and the public sector in areas such as ERP, CRM, HR, payroll data, process migrations, internet security appliances, and B2B marketplaces. He has delivered Just-In-Time logistics and series production systems for customers such as BMW, Volkswagen, and Mercedes.

Andreas completed his engineering degree in Germany and got his first Commodore 64 when he was 12 years old. Originally from Bavaria, Andreas now lives in Sydney, Australia.

About the reviewers

With over two decades of experience in the technology industry across the globe, **Niraj Naidu** is a recognized thought leader in driving large-scale business and digital transformation initiatives and optimizing technology strategies. He has excelled in leading enterprise architecture and engineering practices, delivering value, and stewarding impactful customer outcomes. With expertise in strategic architecture delivery and agile software development, Niraj has implemented enterprise technology solutions that add value and stability to businesses. As a customer-focused technology strategist and leader, he continues to provide strategic guidance to help Fortune 500 organizations realize and maximize the value of their technology investments.

My sincerest thanks to the authors of this well-researched technical book for their invaluable contributions. Their expertise, leadership, meticulous approach, and practical insights have not only created a valuable resource for the technical community but have also elevated the understanding and application of the subjects covered. Their dedication to providing comprehensive and accessible content has made this book an indispensable guide.

John Heaton is an innovator, technologist, and leader who builds human-centered organizations enabled by innovative data-driven technology. John has worked independently and for large multinational organizations in a variety of roles covering technology governance, risk and compliance, enterprise strategy and architecture, big data and analytics, and portfolio and program management from conception through to implementation. John now works within organizations to design, build, and operate businesses focused on driving digital transformations. Recently, he helped develop and implement the digital strategy and operating model launching a successful 100% cloud-native digital bank.

Guillaume Poulet-Mathis is a senior technology executive with over 15 years of experience in enabling product innovation. He currently serves as the director of product engineering at Optus, an Australian tier-1 telecommunication company, where he is responsible for the engineering division building the Optus Living Network, a collection of unique on-demand and real-time network features redefining connection experiences for Optus customers. Guillaume spearheaded several technological innovations, such as cloud-native voice network functions and the introduction of an event-driven microservices platform on Kubernetes, enabling engineers to build products seamlessly spanning digital channels and telco network elements.

I would like to thank the numerous open source communities and advocates who have contributed to making technology more accessible and resilient. It is the contributions of passionate individuals that are improving our ways of working and powering the right level of optimism to solve today's world problems with the power of many.

Sharad Gupta leads the Pre-Sales Solutions Engineering team at UiPath and has previously led similar teams at Pivotal (VMware) and DataStax. Sharad's passion is to help customers implement modern architecture practices for resilient and scalable business operations.

In his past roles, Sharad led and advised on enterprise architecture and integration patterns for Fortune 500 companies. He received his master's in computer engineering from Drexel University and an MBA from Fisher College of Business at The Ohio State University.

With constant curiosity about the world around us, Sharad enjoys engaging in discussions on topics such as process improvement, behavioral economics, business strategy, and business transformation.

Thenna Raj is an outcome-driven technology leader with over 20 years of experience in technology strategy, architecture, and delivery in complex environments across a range of industries, including retail, government, telecommunications, and financial services.

He specializes in delivering strategy and architecture functions across various business brands. Over the course of his career, Thenna has worked across systems engineering and design, business analysis, product management, and consulting roles. Thenna has extensive experience working in an agile environment across various organizations, and his diverse background provides him with the ability to translate strategic objectives into pragmatic outcomes.

Table of Contents

Part 2: Building a Successful Technology Operating Model for Your Organization

5

Building Your Distributed Technology Operating Model   69

6

Your Distributed Technology Operating Model in Action   117

7

Implementing Distributed Cloud and Edge Platforms with Enterprise Open Source Technologies   145

8

Into the Beyond   171

Preface

Speeding up time to market, lowering the **total cost of ownership** (**TCO**), reducing CapEx, enabling self-service, and reducing complexity are important cloud goals; however, the desired outcomes don't always materialize. With edge computing making its way through all industries on top of ongoing journeys to the public cloud (and back), it's vital to share working recipes for organizations to find their preferred way of extracting the most value out of their technology investments.

This book demonstrates a practical way of building a strategy-aligned operating model while considering a variety of related aspects such as culture, leadership, team structures, metrics, intrinsic motivators, team incentives, tenant experience for development and product teams, platform engineering, operations, open source, and technology choices – just to name a few. You'll understand how single, multi, or hybrid cloud architectures, security models, automation, application development, workload deployments, and app modernization can be re-utilized for edge workloads to help you build a secure, yet flexible technology operating model. You will learn how to build a distributed technology operating model using a case study.

By the end of this book, you'll be able to build your own fit-for-purpose operating model for your organization in an open culture way.

Who this book is for

As a cloud architect, solutions architect, DevSecOps or platform engineering lead, program manager, CIO, CTO, or chief digital officer, if you're tasked to lead cloud or edge computing initiatives, create architectures and enterprise capability models, align budgets, or show your board the value of your technology investments, then this book is for you. This book will help you define and build your cloud and edge computing capabilities around a fit-for-purpose technology operating model.

What this book covers

Chapter 1, *Fundamentals for an Operating Model*, looks at the challenges that the journey to the cloud has thrown at organizations and why. It goes through a rich set of examples to look at different ways to define key components of an operating model: the operating model dimensions. It also introduces key terms from engineering and operations and distinguishes between platform and product engineering and SRE. This chapter closes with thoughts on how to construct metrics, teaming, and how Conway's law affects your architecture.

Chapter 2, *Enterprise Technology Landscape Overview*, gives an overview of common enterprise technology landscape components. It distinguishes between systems of innovation, differentiation, and systems of record, and examines the associated change cadence and challenges related to these cadences by expanding the classification from applications to infrastructure. It completes the discussion by looking into the difficulties around the adoption of a standard operating model because of the distinguished traits across that classification.

Chapter 3, Learnings From Bimodal IT's Failure, examines closely why the Gartner Bimodal IT approach never yielded the results expected and extracts the learnings out of it in order to apply them to the distributed technology operating model. It looks at the change cadence differences between mode 1 and mode 2. It closes by highlighting that there is no endorsed bimodal architecture that combines a working approach between mode 1 and mode 2 estates.

Chapter 4, Approaching Your Distributed Future, focuses on the imminent distributed future. It looks at the reasons why the future is distributed and revisits hybrid and multi-cloud definitions while shedding light on specific business and technology reasons why the public or single cloud cannot be a target state for organizations. It spends time on different edge classifications to get a better handle on different viewpoints in light of worthwhile use cases. It closes by looking at emerging trends and external factors such as compliance, mergers, and acquisitions.

Chapter 5, Building Your Distributed Technology Operating Model, explains in detail the building blocks for a distributed operating model across the cloud and edge. It starts off by showing the steps toward the desired outcome in the *Starting at the end* section and introduces an operating model dashboard to track outcomes, **work in process (WIP)**, and dependencies. It presents workshop-leading practices to help lead teams along the forming, storming, norming, and performing life cycle. The chapter also walks through more than 30 dimensions to consider and choose from for the operating model. It also provided numerous suggestions for further reading if you want to dive deeper into any of the research underpinning our recommendations.

Chapter 6, Your Distributed Technology Operating Model in Action, introduces an anonymized real-life use case and walks through how this organization built its distributed technology operating model in a hybrid multi-cloud and edge context.

It walks through the step-by-step process of utilizing already introduced templates and new assets that can be reused for the operating model development process.

Chapter 7, Implementing Distributed Cloud and Edge Platforms with Enterprise Open Source Technologies, walks through an operating model-based platform implementation example. It connects real work architecture, design, and implementation with the previously developed operating model. It shows how requirements and principles from the operating model flow into technology selection and how they map to capabilities.

Chapter 8, Into the Beyond, wraps the book up. It introduces additional aspects such as antifragility, geographically disparate (non-)autonomous operating models; different ways to measure progress; how tech debt, undifferentiated heavy lifting, and open source are connected; gap analysis; and roadmap development. It revisits prioritization and decision-making and introduces a quick way to make the best possible decision with the often limited information available.

To get the most out of this book

With prior knowledge of cloud computing, application development, and edge computing concepts, you will get the most out of this book.

You may appreciate it more if you are in a role such as CxO, senior business/IT leadership, enterprise architecture, or the development and operations teams.

Conventions used

There are a number of text conventions used throughout this book.

Code in text: Indicates code words in text, database table names, folder names, filenames, file extensions, pathnames, dummy URLs, user input, and Twitter handles. Here is an example: "Prakash captured it and followed a similar approach to help the steam team define the transition states for the PTE. Platform.UnifiedPlatform item."

Bold: Indicates a new term, an important word, or words that you see onscreen. For instance, words in menus or dialog boxes appear in bold. Here is an example: "For example, the two dependencies that were identified under **LA.Customer** were **LA.Architecture.Resilience** and **LA.Architecture.Security**."

> **Tips or important notes**
> Appear like this.

Get in touch

Feedback from our readers is always welcome.

General feedback: If you have questions about any aspect of this book, email us at customercare@packtpub.com and mention the book title in the subject of your message.

Errata: Although we have taken every care to ensure the accuracy of our content, mistakes do happen. If you have found a mistake in this book, we would be grateful if you would report this to us. Please visit www.packtpub.com/support/errata and fill in the form.

Piracy: If you come across any illegal copies of our works in any form on the internet, we would be grateful if you would provide us with the location address or website name. Please contact us at copyright@packt.com with a link to the material.

If you are interested in becoming an author: If there is a topic that you have expertise in and you are interested in either writing or contributing to a book, please visit authors.packtpub.com.

Share Your Thoughts

Once you've read *Technology Operating Models for Cloud and Edge*, we'd love to hear your thoughts! Scan the QR code below to go straight to the Amazon review page for this book and share your feedback.

https://packt.link/r/1837631395

Your review is important to us and the tech community and will help us make sure we're delivering excellent quality content.

Download a free PDF copy of this book

Thanks for purchasing this book!

Do you like to read on the go but are unable to carry your print books everywhere? Is your eBook purchase not compatible with the device of your choice?

Don't worry, now with every Packt book you get a DRM-free PDF version of that book at no cost.

Read anywhere, any place, on any device. Search, copy, and paste code from your favorite technical books directly into your application.

The perks don't stop there, you can get exclusive access to discounts, newsletters, and great free content in your inbox daily

Follow these simple steps to get the benefits:

1. Scan the QR code or visit the link below

https://packt.link/free-ebook/9781837631391

2. Submit your proof of purchase
3. That's it! We'll send your free PDF and other benefits to your email directly

Part 1: Enterprise Technology Landscape and Operating Model Challenges

In this part, you will get an introduction to key concepts associated with a cloud operating model, see an overview of the enterprise technology landscape, and learn how Gartner's pace layered architecture can be used to classify applications and infrastructure based on key characteristics. It also covers why previously hyped concepts such as Bimodal IT didn't take off, distills knowledge about its limitations, and explains the reasons why the future will be distributed for organizations.

This part has the following chapters:

- *Chapter 1, Fundamentals of the Cloud Operating Model*
- *Chapter 2, Enterprise Technology Landscape Overview*
- *Chapter 3, Addressing Diverse Technology Landscapes with Bimodal IT and Its Limitations*
- *Chapter 4, Approaching Your Distributed Future*

1
Fundamentals for an Operating Model

In this first chapter, we will set the context for the remainder of this book and clarify some terms. In an industry full of buzzwords and ambiguity, we felt that level-setting and providing context deserves a chapter of its own. This will help you understand the rest of this book and communicate better based on agreed-upon terms and associated semantics.

In this chapter, you will learn about various concepts related to your strategy, goals, mission, vision, and objectives by looking at different frameworks. Additionally, you will explore different definitions of operating models and extract essential insights from them. This chapter will also delve into the significance of culture in creating high-performing organizations and teams, along with other important topics, such as capability and platform engineering. Overall, this chapter aims to equip you with a comprehensive understanding of the key elements that are necessary for successful organizational management.

Our goal is to provide you with answers to the following inquiries:

- What exactly is an operating model and what is its significance?
- How does an operating model relate to strategy?
- What frameworks are available concerning operating models?
- How can you effectively implement an operating model in your organization?

By addressing these questions, we hope to equip you with a comprehensive understanding of operating models and their role in organizational success. By the end of this chapter, you will know what an operating model is and learned about business operating models and how we can translate them into technology operating models.

Why this book?

Simon Oliver Sinek is a British-born American author and inspirational speaker. He is the author of five books, including *Start With Why and The Infinite Game*. Let's follow Simon Sinek's advice and Start with Why. Every organization's "why to use the cloud" could be subtly different. However, there are common ones such as CapEx, OpEx, or risk reduction, as well as faster time to market. Getting the desired **return on investment** (**ROI**) out of moving to the public cloud is not as easy as it looked when the hyperscalers sales team did their presentations. We are not going to cite percentages here, such as "X% of all cloud migrations fail." First, we don't have a consistent and overarching definition of "failure," nor a specific one for each case. Second, every "failing" initiative has brought learnings with it. And third, we need to appreciate these first movers because we benefit from their learnings. However, research suggests that a significant number of cloud migrations do not go as smoothly as initially anticipated.

The reasons for limited cloud success can vary from organization to organization. Here are some examples:

- Unclear direction-setting attempts like 'cloud first' left teams unsure of their future or what to do.

- Moving to the Cloud is mistaken for innovation.

- Goals to move large amounts of applications to the public cloud in an established enterprise without significant change management are not realistic.

- Companies experienced massive margin pressure because the cost of existing infrastructure didn't disappear as quickly as anticipated.

- Operational expenditure (OpEx) and bill shock from the hyperscalers occurred due to the use of popular high double-digit gross-margin cloud services (compared to single-digit gross margins for owned hardware).

- Lift and shift shortcuts do not leverage Cloud on-demand scale-out/in features and, as a result, do not lead to the desired business demand-aligned pricing.

- Data mobility needs and associated egress costs were not taken into account.

- Marketing metrics driving wrong behaviors: "30 apps in 30 days" slogans sound good; however, the objectives were wrongly focused on the number of applications moved to the cloud instead of creating and deploying cloud-ready, cloud-optimized, and cloud-native applications that generate a return on investment (ROI).

- Seeing the public cloud as the ultimate target state: Shortcomings of architecting for an edge-inclusive distributed future led to low-ROI efforts of moving non-fit workloads to a public cloud environment.

- Unrealistic expectations leading to unrealistic timelines: For organizations with hundreds or thousands of applications, it is unrealistic to expect to move quickly, assuming application refactoring will happen simultaneously across hundreds of applications that can be worked on at the same time.

- Focus on technology instead of transitioning people, processes, and culture to enable cloud-ROI through microservices, containers, DevSecOps, GitOps, Agile delivery, and self-service.

- Data sovereignty, residency, and data concentration risk need mitigation through resilient architectures, which increase efforts and timelines.

- New cybersecurity threat vectors - the need for a consistent security posture across heterogeneous infrastructure footprints is more complex than traditional perimeter-based security.

- Multiple proprietary public cloud implementations required reinventing the wheel and offered little reuse.

- Even DIY private cloud builds, where the team had intrinsic business knowledge but followed a 'build it and they will come' approach, had limited success with adoption.

The main root cause of these challenges often lies within "organic – just do it" cloud journeys and the fact that these journeys didn't start with an "operating model first" approach. Now, if we take all of these challenges and add edge computing to them, the chances of success become even less likely. The added number of edge locations, hardware, platforms, deployments, applications, data, and security requirements amplify the complexity. This shows what edge computing can potentially bring in terms of these challenges – it calls for a different, less fanboy (and girl!)-ish approach. And that's why we recommend an operating model first approach.

The recommendation is to not start with logos that we want on our CVs, or our love for technology – not even a specific use case. We must start with the operating model. Even though operating models don't last forever, they are usually longer lived than strategies and hence a good fit to use as an anchor. Operating models – if done well – are the glue between a company's strategy and the people that make this strategy come alive during its execution.

Defining the cloud and the edge – hybrid cloud, multi-cloud, plus near and far edge

Cloud computing is a model for delivering computing resources, such as servers, storage, databases, and other services, over the internet. In cloud computing, users can access and use these resources on demand, without having to own and maintain their computing infrastructure.

Cloud computing is typically provided by third-party cloud service providers who manage and maintain the underlying hardware and software infrastructure, as well as provide the necessary network connectivity, security, and support services.

There are three main types of cloud computing services:

- **Infrastructure-as-a-Service (IaaS):** This provides users with access to computing infrastructure such as virtual machines, storage, and networking resources

- **Platform-as-a-Service (PaaS)**: This provides users with a platform for developing, running, and managing applications without having to worry about the underlying infrastructure

- **Software-as-a-Service (SaaS)**: This provides users with access to software applications that are hosted and maintained by the cloud provider

Hybrid cloud is a computing environment that combines both private and public cloud infrastructures. With hybrid cloud, organizations can use both on-premises (private cloud) and public cloud-based resources. Co-location providers such as Equinix count as on-premises and/or private clouds.

This approach enables organizations to take advantage of the scalability and cost-effectiveness of public cloud resources while maintaining sovereignty over their sensitive data and applications through private cloud resources. Organizations usually run their non-cloud-ready applications using 24/7-always on or monolithic core systems of records-type applications such as non-SaaS ERP, CRM, finance, and HR systems, such as on-premises ones.

Multi-cloud refers to a cloud computing environment that involves using multiple (public) cloud service providers to host different parts of an organization's computing infrastructure or workloads. In other words, instead of relying on a single cloud provider, a multi-cloud strategy involves using multiple cloud providers in a coordinated manner.

Edge computing, on the other hand, involves processing data closer to where it is generated, rather than in a centralized cloud environment. It refers to the use of decentralized computing resources that are located at or near the edge of a network, rather than in a centralized data center. This can improve the speed and efficiency of data processing and reduce latency. Edge computing typically involves small, distributed computing resources located at the edge of the network, such as sensors, small form factor compute devices, robots, or edge servers. Depending on the distance from the user or data center, edge computing can be further categorized into near and far.

The associated benefits allow organizations to process data closer to the source of the data, which can reduce latency and improve the performance of applications and services and reduce network design complexity. By including edge computing in a hybrid and multi-cloud model, organizations can take advantage of the flexibility and scalability of the cloud, while also being able to process data in real time at the edge of the network. This ultimately enables organizations to address requirements and execute use cases that were not possible before.

Together, cloud and edge computing create a comprehensive computing environment that combines the benefits of both approaches. For example, an organization might use a hybrid cloud to store and manage its data and run both cloud-native and monolithic applications while using edge computing to process data generated by IoT or other edge devices in real time.

Setting the organizational context – strategy, culture, capabilities, operating model, and more

In today's world, acronyms are everywhere! A **three-letter acronym** (TLA) can mean different things: **OSS** could mean **Open Source Software** or, in a telco context, **Operational Support System**, **Open Sound System** (Unix), or something else. And as you know, there are plenty of other examples out there. So, you understand why it's important to set the context in your organization too. Here's an example from a previous employer of mine: *PS* stood for *professional services*, as well as *pre-sales*. You can imagine how many unnecessarily confusing situations that caused. So, my recommendation is to kill not Bill, but ambiguity. This is worth it. The practices we will introduce in *Chapter 5*, *Building Your Distributed Technology Operating Model*, will help you achieve this.

In this first chapter, we will provide some context and a description of what we mean by the terms and terminology we use. We are slaying buzzwords right here, right now. Let's get to it.

Strategy

A **strategy** is an integrated set of choices an organization makes, without really knowing if they work. A strategy is a set of hypotheses that you think will help you win on a playing field of your choosing. So, a strategy is based on a theory. That theory should be coherent and executable by the people in your organization right now. What winning looks like is defined by the strategic goals you define. Ideally, a strategy is communicated to your colleagues so that, as a team, you can all pull in the same direction. As an example, regarding the playing field of your choosing the Amazon bookstore decided to extend their playing field from an online bookstore. First, it was to become "The Everything Store" and then a public cloud service provider.

To clarify how to develop and set an operating model in the context of our strategy and goals, we can utilize an existing framework from the **Business Motivation Model** (BMM). The OMG Group's BMM includes the **Means to End** framework. Means is the action plan, while End is the desired result or aspiration. You can study the BMM meta model via the link provided in the *Further reading* section and learn about the entities and relationships in more detail, but it's not necessary to do so for this book.

The Means to End framework aims to put concepts such as *Mission*, *Vision*, *Strategies*, *Goals*, *Objectives*, and *Tactics* into context and defines a common language. A common language is a very powerful enabler. The information exchange and hence learning and understanding that occurs across teams, even from within the same organization because of that common language, is phenomenal.

I've run many workshops where this simple framework created a lot of clarity for the customer's team, which is why I recommend it. I also added a link in the *Further reading* section in case you want to facilitate a workshop yourself.

Introducing the Means to End framework (see *Figure 1.1*) to workshop participants is an effective way to link their vision (for example, we want to be a digital bank with a brick-and-mortar experience) and mission (we prioritize building out our digital CX), as well as their strategy to goals, and distinguish between strategic goals (for example, grow assets under management beyond $80 billion for a bank) and associated tactical objectives (for example, automate 100% of the loan origination process). At this point, a valid tactic could be to fund a project that digitizes the enter loan origination process and the associated strategy to build out straight-through processing for all asset-related customer touchpoints:

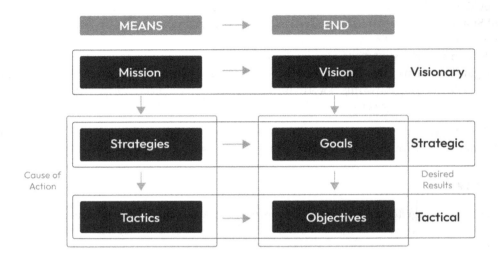

Figure 1.1 – The Means to End framework

So, even though we ultimately talk about strategy, let's take a quick detour to see how strategy is connected to the other elements you encounter in your organization. This will be useful later when we define the "success criteria" – that is, our goals and objectives – as we move toward our target state operating model. Let's quickly go through the different elements of the framework and give some examples of what we mean by that:

- A **vision** represents an organization's future and is answering the question of who we are going to be in 2, 3, or 5 years from now.

- The **mission** is the means to achieve the vision (end) and sets the direction by stating what organizations do daily to achieve our vision.

- **Goals** are connected to the vision because the goals that have been set need to align with your organization's vision. Because we are in the "strategic" layer, goals are strategic and hence answer the question, "What strategic goals do we need to hit to make this vision a reality?" Goals are longer-term but should be narrow enough and have qualitative definitions so that objectives can be created for them.

- **Strategies** are the means we choose to achieve our strategic goals (end). In this layer, we are figuring out what high-level approaches (programs of work, products, or services) and hypotheses are being funded to achieve our goals. Strategies are usually broad in scope and long-term compared to tactics. Think of a program and product instead of a project. As you can see, the mission informs the strategies.

- **Objectives** are steps toward a goal. They should be specific and of a qualitative nature with an end date to ascertain whether the goal has been reached or not. Objectives need to be linked to strategic goals; otherwise, you need to ask yourself: Why am I doing this?

- Finally, we have **tactics**. What tactical projects or tasks (means) do we employ to achieve our objectives (end)? Tactics are short-term and narrower in scope – think project or feature rather than program or product. Strategies inform the tactics, and the tactical objectives should support the strategic goals. To summarize, every objective you achieve brings you closer to reaching the associated strategic goal.

But I need to utter a word of warning: you can run into difficulties distinguishing between the strategic and tactical layers at times during workshops. A pro tip is to keep in mind that strategies and goals are usually longer-term and broader in scope. Tactics, on the other hand, are shorter-term and narrower in scope. The following diagram shows how you can outline the context of your strategy on a single page:

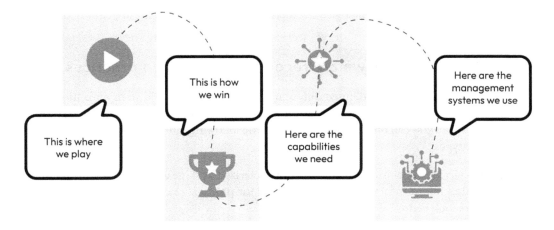

Figure 1.2 – Visualizing the big picture – a single-page overview of strategic context

And how is this related to the cloud operating model I came here for, you might ask? Great question! A business operating model needs to support the company's strategy. For example, if you want to grow revenue and venture into customer segments by bringing new features or products faster to market while utilizing Agile and microservices but your IT operating model is set up to stabilize your systems of records, you might end up wasting lots of effort and money. So, the operating model needs to align if you want to be efficient and effective.

And the same is true for your cloud operating model and strategy. If you want to reduce your time to market, attract and retain talent, reduce OpEx, be more innovative, reduce tech debt, or improve your bottom or top line, then you need to do more than just select a hyperscaler to run on.

In short, your operating model needs to encompass things such as funding (project or product?), team setup (Conway's law or Dunbar's number?), platform (where and what to abstract?), cultural practices (Open Practice Library and/or DevSecOps?), and much more. This is the core of your cloud operating model. We will cover this in more detail in *Chapter 5*.

Capability

Capability is, in general, defined as the power or ability to do something. **The Open Group Architecture Framework** (**TOGAF**) defines it as *an ability that an organization, person, or system possesses.* Product features can sometimes be referred to as capabilities. However, product or system features are not what we mean in this book when we say capability. We mean the organizational capability that enterprise architects refer to when people, processes, and technology are in place and form the ability of an organization to do something. Such a capability could be product marketing, processing an insurance claim, or employing DevSecOps practices. If an enterprise has any of those capabilities, then the people, processes, and technologies are in place. And that's what we mean in this book when we refer to "capability."

About culture and why we are recommending open practices

Sociologist Dr Ron Westrum has defined different cultural typologies within organizations and his research has shown how culture affects performance. The different typologies and the "cultural features" or behaviors you can observe are depicted in the following table, but to summarize, there are three culture types:

- A pathological (power-oriented) culture is characterized by large amounts of fear and threat. People often hoard information or withhold it for political reasons or distort it to make themselves look better.

- In bureaucratic (rule-oriented) cultures, organizations protect departments. People in the department want to maintain their "turf," insist on their own rules, and generally do things by the book – *their* book.

- Generative (performance-oriented) cultures help organizations focus on their missions and goals. A generative culture allows people to openly ask, "How do we accomplish our goal?" Everything is subordinated to good performance and doing what needs to be done to get things done.

These are all specific features of Westrum's cultural categorization. Overall, the *mission and goals* take precedence in a *generative or performance-oriented* culture. And that's certainly what you want if you want a far-reaching concept such as an operating model to come alive.

If we hone in on specific features within that cultural typology, Westrum found that a performance-oriented culture displays behaviors such as high cooperation, with novelty being implemented/welcomed and information is freely available. And that, beautiful people, is exactly what open source is all about. Here is a table outlining the behaviors that can typically be observed in each of the cultural types:

Ron Westrum

Pathological Power Orientated	Bureaucratic Rule Orientated	Generative Performance Orientated
Low cooperation	Medium cooperation	High cooperation
Novelty Crushed	Novelty is a troublemaker	Novelty implemented
Information Hoarding	Selective Information Flow	Information freely available

Figure 1.3 – Westrum's cultural typology and associated behaviors observed

Git repositories contain freely available information where anyone is welcome as a contributor, new ideas are discussed, and new features and bug fixes are implemented and merged via pull requests. So, how do we transfer this code-centric approach to workshops, status meetings, brainstorming and discovery sessions, strategy discussions, and finally into building a cloud operating model?

The answer is *open practices*.

Westrum also observed that organizational culture affects how information moves within an organization. Westrum provides three characteristics of good information:

- It provides answers to the questions that the receiver needs answering
- It is timely
- It is presented in such a way that it can be effectively used by the receiver

Open practices enable and create "good information" through globally proven practices that are freely available in an open source manner to everyone who needs to facilitate workshops, drive consensus or innovation, discussions, or brainstorm sessions through the Open Practice Library. The Open Practice Library uses a modified Mobius Loop to sort practices into four categories: Foundation, Discovery, Delivery, and Options. A **Mobius Loop** is a horizontal figure of 8 representing an infinity loop sitting on top of Foundation practices that iterates through Discovery, Delivery, and Options indefinitely.

For each part of creating the operating model, I will recommend specific exercises that you can run with your team and stakeholders. This will help you establish an open culture, utilize the wisdom of the crowd (as opposed to following a detrimental **Highest Paid Person's Opinion** (**HiPPO**) approach, and create a sense of ownership in your organization.

And that's because people – especially the people who do the work – have a say instead of just being a recipient of decisions – in our case, decisions around a (new) cloud operating model. Being a passive recipient is – in my experience – more likely to create change resistance than excitement and ownership. And a sense of ownership is what we need if we want to create and, subsequently, iteratively improve our cloud operating model.

Operating model

First of all, an operating model and an operations model are two different things. When we talk to organizations, we often start with this statement as it clears up potential misunderstandings right from the start. They are related, but different. There is no single place to go look up the definition of what a best-practice operating model is. And worse, each management consultancy has different definitions, usually as a competitive differentiator. Let's have a quick look at the definition of an operating model.

An operating model is a plan or system that outlines how an organization will function and achieve its goals while delivering value for its customers. It defines the processes, resources, structures, management approaches, and systems that are needed to support the operations of the business. It also outlines the roles and responsibilities of different teams and individuals within the organization, as well as the relationships between them. An operating model is an important tool for ensuring that an organization can operate efficiently and effectively, and it is used to support strategy and decision-making.

An operating model is divided into different categories or dimensions in a divide-and-conquer kind of fashion. This helps us focus on specific aspects within the vast spectrum of things to consider while running an organization.

Even though our ultimate focus in this book is on distributed technology operating models in the context of hybrid multi-cloud and the edge, it's good to start by looking at more general business operating models first and how different subject matter experts define them. The learnings and observations you gain will make it easier for you to work on a cloud operating model and its associated dimensions.

Operating model dimensions

The dimensions of an operating model help you categorize specific topics so that you can focus on and hone in on them. As I said previously, there is no set of globally agreed-upon dimensions. Let's take a look at what some expert management consultancies suggest to help you categorize the problem space.

Accenture

Accenture, in their proposal of a "resilient operating model," uses the following categories:

- Agile governance organizes the workforce and transitions the way work is done, promoting a culture of experimentation and innovation, faster decision-making, and approval cycles regarding how performance is measured.

- Taking a two-pronged technology approach is about investing in new and sweating existing technology assets on an incremental and ongoing basis to continuously evolve their capabilities.

- Configure and reconfigure talks about creating squads, pods, or cells that operate like discrete businesses within the organization to help boost agility and responsiveness for specific products and services.

- Invigorating the ecosystem includes reevaluating partners, channels, and services and developing capabilities to help achieve long-term objectives such as driving innovation, developing new products and services faster, entering new markets, or being more agile.

- Decision-making at the edges is about combining a real-time data access capability with the cultural shift to empower employees to make decisions based on what they see in the data.

- Reskill, reskill, reskill suggests what most of us technologists know already: we are never done learning. It's about building a culture of continuous adaptation and building and rebuilding the skills of employees, including the skills required to work with the latest technologies to create a *human + machine* mindset.

KPMG

Now, let's look at what **KPMG** (one of the Big Four accounting organizations) has to say. KPMG calls their operating model the **target operating model** (**TOM**), which is a concept we will also look at for our technology operating model later in this book. Here, we map out what the target state is in certain areas – for example, product-based budgeting – and then assess where we are to find out how to get to the target. We'll cover this in more detail later in this book.

So, let's go back to our friends at KPMG. The KPMG TOM dimensions are as follows:

- Functional process

- People – who does what, reporting lines, required skills, roles, and responsibilities

- A service delivery model, such as a shared service center, **Center of Excellence** (**CoE**), outsourcing, and service delivery optimization

- Technology relates to applications and integrations that enable processes with cloud architects, integrations, conversions, and test scripts

- Performance insights and data talks about the what and how of reporting, associated information requirement, and KPI frameworks to optimize decision making

- Governance to cater for oversight and define risks and controls, segregation of duty, and access rules and policies

The context of the KPMG TOM is enterprise transformation and has a strong process focus. The KPMG TOM also comes with blueprints, including process maps. In one of my last enterprise transformation programs, we had several larger management consultancies involved but for process mapping, we decided against the use of any proprietary process maps. We went with APQC and their industry frameworks instead as a baseline.

Forrester

So, what's **Forrester** saying? The latest edition is about an operating model centered around "customer obsession." So, the customer operating model talks about the following dimensions:

- Strategy
- Vision
- Culture
- Performance
- Corporate values
- Motivation

Then, it dives into sub-dimensions such as accountability and compliance. Only a few layers down, we get to more tangible topics such as operating units, location, reporting lines, infrastructure, applications, people, data, and processes, and finally to the customer journey, customer experience, product and service offerings, and value propositions – perhaps too many things to be practicable. But it's not easy to consolidate so many important aspects into the right amount and the right dimensions, as we will see later. The Forrester context is the IT operating model but with a focus on customers. We dig customer focus. A lot.

McKinsey

Lastly, before we wrap things up, let's have a look at *McKinsey*, which has three high-level dimensions called *People*, *Processes*, and *Structure*, and then lower-level sub-dimensions all centered around *Strategy*:

- People:
  - Informal networks
  - Culture
  - Talent and skills
  - Workforce planning
- Structure:
  - Roles and responsibilities

- "Boxes" and "lines"
- Boundaries and location
- Governance

- Processes:

  - Process design and decisions
  - Performance management
  - Systems and technology
  - Linkages

In summary, regardless of what different names different experts use, it's safe to say that for an organization to deliver value to its customers, the following operating model dimensions must exist under one name or another:

- Process
- Organization
- Location
- Information
- Supplier
- Management systems

All these operating model dimensions should contribute to an organization's value chain. If this sounds a bit like a business model now, then have a look at *Figure 1.4*:

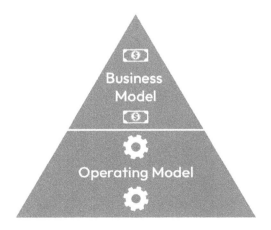

Figure 1.4 – Business versus operating model

An operating model and a business model are two related but distinct concepts:

- A **business model** is a high-level framework that describes how an organization creates value for its customers. It defines the value proposition, target market, revenue model, and cost structure, and outlines how these elements work together to create a sustainable and profitable business.

- An **operating model** is a more detailed framework that outlines how organizations execute their business model. It defines the organization's strategy, structure, processes, people, technology, and governance, and outlines how these elements work together to deliver value to customers and stakeholders.

In other words, while a business model describes what an organization does and how it generates revenue, an operating model describes how it manages resources, processes, and activities.

Our little excursion into the world of operating models also showed that it can be quite frustrating if you are looking for *the one and only* operating model, simply because it doesn't exist. And that is true for general business and organizational operating models as much as it is for cloud operating models.

If we look at it differently, it's quite liberating and reassuring that we have the freedom to create and employ the best-fit operating model for ourselves. And the best thing is that we can involve our peers, teams, managers, and direct and indirect reports to ensure we create something that is a) fit for purpose and b) that people feel a sense of ownership with. And if people have a sense of ownership, we have a chance of getting our distributed cloud operating model adopted.

And just to finish up, in case you are interested in what a meta-model looks like and to be a bit more scientific, the operating model can extend the business motivation meta-model to show and describe the relationship between the different entities that play a role in defining an organizational and, ultimately, hybrid cloud operating model:

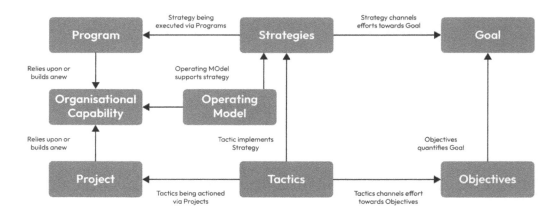

Figure 1.5 – Business motivation model with our operating model extension

As you can see, the operating model entity is connected to the strategy. This is because we said it needs to support the strategy – remember the cloud-native microservices versus mainframe example from earlier? The operating model entity is also related to organizational capability. This is because if the organization doesn't have the required capabilities – that is, people, processes, and technology – then it can't effectively and efficiently operate and execute projects or larger programs of work. This, in turn, means that organizations that need a DevSecOps capability to reach their distributed cloud operating model have to build it.

Operating models for hybrid cloud and the edge

Earlier, we looked at business operating models. We could now go on and do the same exercise as in the previous section and see what global system integrators and consultancy companies have in terms of cloud operating models so that we can inform our distributed technology operating model (including the edge), but we will find the same thing is true here.

And while hybrid or multi-cloud is a reality and a necessity for most organizations, we want to go further. We want to incorporate edge computing. Edge computing is becoming pervasive across all industries. There are obvious use cases in this category, such as self-driving cars or mobile phone towers as part of the telco provider's edge **radio access network** (**RAN**), as well as operational edge scenarios for monitoring manufacturing plants.

A hybrid cloud and edge operating model is a type of hybrid cloud operating model. It involves the use of both public cloud services and a private cloud or on-premises infrastructure, as well as edge computing locations.

As part of a hybrid cloud and edge operating model, an organization can choose to run certain workloads on the public cloud, others on the private cloud or on-premises infrastructure, and still others on edge computing resources, because the operating model dimensions cater to this scenario.

A hybrid cloud and edge operating model allows organizations to adjust their computing and data deployments to the specific needs of their users and customers. This is because the relevant distributed security and compliance posture management, development and operational capabilities, architecture, funding, and skills are available to make use of the most appropriate resources for each workload to achieve the desired speed, flexibility, scalability, and cost-effectiveness.

Hybrid cloud computing services provide on-demand access to computing resources such as storage, networking, and processing power over a WAN. This allows the organization to request resources as needed, pay for only the resources it uses, and benefit from the flexibility, agility, and cost-efficiency of the elasticity (scale-out/scale-in) of the cloud. However, this requires suitable workloads and a technology operating model that is ready to leverage this opportunity.

Distributed (as opposed to centralized and single cloud) is key here. In the context of hybrid cloud and edge computing, "distributed" refers to deploying computing resources across multiple locations, including both on-premises data centers, far or near edge locations, and public cloud-based environments.

The ground zero of a hybrid cloud and edge operating model includes the following elements:

- **Distributed architecture**: The design and layout of an organization's distributed cloud and edge-based systems and services, including how they are connected and interact with one another

- **Distributed infrastructure**: The hardware, software, and networking resources that an organization uses to host its applications and data in a distributed manner, leveraging the benefits of cloud and edge computing and other distributed technologies

- **Distributed security**: The measures and controls that an organization puts in place to protect its distributed cloud and edge assets from threats and vulnerabilities

- **Distributed management**: Control plane processes and tools that an organization uses to monitor, optimize, and maintain its distributed cloud-based systems and services

Proactively creating a fit-for-purpose cloud and edge operating model enables organizations to ready themselves for the distributed future. Doing so cultivates the potential to outdo the competition. Outdoing the competition can have many forms: being more agile, quicker to market, more cost-effective, working with less risk, more secure, and so on.

We can expand on the infrastructure layer and then the operating model for the distributed future using a set of practices, processes, and technologies to operate and manage its applications, data, security, and compliance posture on top of distributed infrastructure. A "distributed infrastructure" simply means a mix of on-premises, private, and public cloud, as well as edge locations.

We believe – and research suggests – that the future will be highly distributed and that organizations are likely to rely more heavily on combinations of cloud computing, edge computing, and other distributed approaches to deliver their products and services related to customer experience. This may involve leveraging the scalability, reliability, and cost-efficiency of a private and public cloud, as well as the low latency aspects of edge computing.

So, why do we need an operating model? Because an operating model allows us to look at different aspects of our execution, which enables strategy and business model alignment in a reusable manner with a focus on "moving target" outcomes. We shall call those aspects dimensions henceforth.

While organizational operating models balance integration versus standardization or localization against globalization or centralization versus decentralization, the technology operating model balances the same tensions by providing innovative freedom versus operational excellence. More details on this will be provided in *Chapter 5*, and *Chapter 6*.

Before we dive into the details of how you can craft your own fit-for-purpose distributed technology operating model, perhaps you should start thinking about what aspects or dimensions you would choose if it was all up to you.

Because we are still in the "foundations" part of this book, next, we will define a few more terms to provide clarity for later sections of this book.

Engineering and operations

Just a word of warning: it can get blurry!

Software product engineering and operations (including support and maintenance) are related but distinct activities within the field of software development. In the early days, at least. Then, Dev{X} Ops came along, and it got harder to talk to one another about this topic.

For the sake of this book, we'll define three things:

- A blurry universal truth about engineering and operations
- Platform engineering (where the buzzword SRE lives)
- Product engineering (where DevOps is at home)

Platform and product engineering both contain operations aspects in their respective fields:

- Platform engineering owns platform operations, support, and maintenance
- Product engineering owns product operations, support, and maintenance

This clarification is important because it helps clarify practices such as DevOps, DevSecOps, and **Site or Service Reliability Engineering** (**SRE**), as examples. In modern organizations, product and platform teams have a product manager, backlog, and practices. So, for example, the platform team could employ an SRE approach, work with their internal customers (the product teams) on defining **service-level agreements** (**SLAs**) and **service-level objectives** (**SLOs**), and collaborate while selecting the most appropriate **service-level indicators** (**SLIs**), which indicate potential **customer experience** (**CX**) impact and hence help raise alarms before CX is impacted.

The dependent product teams use this platform as a service to build their products and service offerings on top and might use different tooling or different approaches such as DevSecOps. Some teams might use a GitOps approach, while others might not.

And this is where the operating model comes in. In the platform and product team example we provided, there were questions about team structure, team collaboration modes, funding, location, and practices. These are all questions that the operating model provides guidance with.

In general, software product engineering focuses on designing and developing new software products, while support, maintenance, and operations focus on maintaining and supporting existing products. However, there is potentially an overlap between the two activities, particularly in organizations where software product engineers are also involved in maintaining and supporting the products they develop (DevOps). We just saw an example of this. And that's why I called it blurry.

The takeaway is that while there is a "best practice" definition, it's always recommended to clarify the actual implementation of engineering practices involved in delivering products and services to production.

Platforms

Despite sometimes not being liked because they introduce an additional "abstraction layer," platforms are generally popular because they address different aforementioned aspects:

- Allow sovereign use and implementation of infrastructure, software, services, and data

- Allow for consistent and SLA-confirmative CX design

- Allow for the required industry-specific compliance and security stance

- Shift security and compliance left without additional burden for developers and product teams

- Allow for a heightened level of reuse and sharing of tools, processes, and digital assets, such as patterns

- Allow the cognitive load of the aforementioned product development teams to be minimized

Having a dedicated platform team to focus on the increasing productivity of product teams, especially in a distributed technology context, is something organizations need to consider. If not an organically grown mess, technical debt and undifferentiated heavy lifting through different tools, services, products, processes, and skills are almost certainly guaranteed, especially in the context of hybrid multi-cloud and the edge.

A platform is a foundation for self-service and a consistent CX. It makes skills, documentation, and processes reusable across both the platform team and the product development teams. The trick is to find the right abstraction layer for compute, network, storage, and databases.

Guiding principles and guardrails

Guiding principles and **guardrails** are both used to provide guidance and structure for decision-making and behavior. However, they serve slightly different purposes.

Guiding principles are softer and found as statements that describe the values and beliefs that guide decision-making and actions within an organization. They help establish a shared understanding of the organization's goals and priorities and provide a framework for making decisions that align with those goals. For example, a guiding principle might be "employee safety first," or "design for workload portability."

Guardrails, on the other hand, are specific and hard guidelines or constraints that are put in place to help ensure that actions and decisions are aligned with the guiding principles. They serve as boundaries that help prevent actions that could lead to negative outcomes or that conflict with the organization's goals. For example, a guardrail might be a policy that requires all new container-based applications to be built from a certified-based container base image, and require security scanning and a signature before being deployed.

In essence, guiding principles are the overarching values and beliefs that guide an organization, while guardrails are the specific rules and policies that help ensure that those values and beliefs are put into action. Together, they provide a framework for decision-making and action that aligns with the organization's goals and priorities.

Being antifragile to change

Robust is out. Antifragile is in. Resilience is a way better mindset than robustness. You don't know what type of storm is coming next, so there's no need to put all you got into an earthquake-safe nuclear power plant if the next thing hitting you is a tsunami. You can disagree, but that's what we believe in. Embracing change is so fundamental that it permeates all operating model dimensions.

In terms of antifragility, there is nothing more antifragile than open source software. Even COVID couldn't stop open source software development based on the distributed nature and cultural aspects of information sharing and high collaboration. Antifragile is more than resilient. Antifragile "things" get better and stronger when exposed to stressors and change, based on the old saying "What doesn't kill you makes you stronger." And that's what the open source community has already proven for many decades. If new challenges emerge, then new projects emerge, and after a while of trying out the best ideas and concepts, these new projects and people converge into only a few and help combine efforts to make those solutions enterprise-ready. And that's antifragile, just like Nassim Nicholas Taleb likes it.

Metrics

I am sure there are really good metrics and really dumb metrics. And even if those metrics appear dumb or smart at the outset, we don't know unless we know how and what they are used for and can see if the metrics are driving the right behaviors. Comparing different teams by comparing user story points is wrong management behavior. As a general rule, though, metrics should always be balanced; otherwise, the system can be rigged. I guess we all had those IT service hotlines at some stage in our professional lives that closed the tickets before the issue got resolved. That is driven by metrics that look only at ticket open times instead of balancing things out with issues resolved. The **DevOps Research and Assessment (DORA)** metrics show this nicely: they balance speed with stability. Speed is measured by deployment frequency and lead time to change, while stability is measured by change failure rate and **Mean Time to Repair (MTTR)**.

On teaming

We believe in setting up long-lived teams. Ample research has been conducted. A short but great read is the book *Team Topologies* if you want to dive deeper into this topic. For example, if you look into Dunbar's number, then you will find that there is a maximum team size. Above that maximum, smaller sub-groupings will appear because we humans can't develop trust among a large number of teammates.

Secondly, each team has been shown to go through different phases: *Forming*, *Storming*, *Norming*, and *Performing*. Knowing the last two phases alone shows that you should not allow short-lived feature teams to be the norm in your organization.

Then, on an individual basis, you can study the results around the intrinsic motivational factors of the knowledge worker, such as autonomy, mastery, and purpose. In an ideal world, your operating model will balance all this out.

On architecture

Conway's law suggests that systems are built according to the existing organizational structure. So, it seems to be a good idea to define an architecture first and then set up a team structure. In nearly all organizations we know, it's done the other way around.

As you can see, there's a lot to do.

So, let's get into it.

Summary

In this chapter, we set the context and defined foundational terms that are important for the following chapters. We covered why it's important to have or even start with an operating model because the cloud and the edge will create too many organic complexities and waste otherwise. We covered fundamental terms such as strategy, teaming, and capability, touched on engineering and operations, and introduced research on culture and how it is connected to an organization's performance. Finally, we covered metrics and team setup and why it affects the architecture.

We are now all systems go to dive into the next chapter, which will provide an overview of the enterprise technology landscape.

Further reading

This section contains links to information that was presented in this chapter so that you can dive deeper into any topics of interest:

- AWS gross margins: `https://www.cnbc.com/2021/09/05/how-amazon-web-services-makes-money-estimated-margins-by-service.html`
- Margin pressures due to the cloud: `https://a16z.com/2021/05/27/cost-of-cloud-paradox-market-cap-cloud-lifecycle-scale-growth-repatriation-optimization/`
- Different operating model definitions:
  - `https://advisory.kpmg.us/articles/2021/kpmg-target-operating-model.html`

- https://www.bain.com/insights/winning-operating-models/
- https://www.accenture.com/us-en/blogs/business-functions-blog/unlock-greater-value-data-driven-operating-model
- https://www.forrester.com/what-it-means/ep258-it-operating-model/
- https://www.accenture.com/content/dam/accenture/final/a-com-migration/pdf/pdf-175/accenture-strategy-resilient-operating-model-pov.pdf#zoom=40
- https://on24static.akamaized.net/event/33/36/01/6/rt/1/documents/resourceList1621516932018/forresteritoperatingmodelwebinar1629744897098.pdf
- https://www.mckinsey.com/capabilities/mckinsey-digital/our-insights/how-to-start-building-your-next-generation-operating-model
- https://teamtopologies.com/book
- https://www.mckinsey.com/capabilities/mckinsey-digital/our-insights/building-a-cloud-ready-operating-model-for-agility-and-resiliency

- Business Motivation Model: https://www.omg.org/spec/BMM/1.3
- Strategy versus planning: https://hbr.org/2022/12/hbrs-most-watched-videos-of-2022
- Facilitating a strategy workshop: https://openpracticelibrary.com/practice/means-to-end/
- APQCs multi and cross-industry process maps: https://www.apqc.org/resource-library
- Intrinsic motivations for knowledge workers:

  - https://www.gse.harvard.edu/news/uk/16/09/intrinsically-motivated
  - https://www.smartcompany.com.au/people-human-resources/leadership/how-to-motivate-your-staff-top-tips-from-daniel-pink/

- Site/service reliability engineering: https://learn.microsoft.com/en-us/azure/site-reliability-engineering/resources/books

- Capability definition: `https://pubs.opengroup.org/architecture/togaf91-doc/arch/chap03.html#:~:text=by%20an%20organization.-,3.26%20Capability,customer%20contact%2C%20or%20outbound%20telemarketing`

- The Everything Store: `https://www.bookdepository.com/The-Everything-Store/9780316377553`

- The future will be distributed research example: `https://www.forrester.com/blogs/predictions-2021-edge-computing-hits-an-inflection-point/`

- Platforms: `https://tag-app-delivery.cncf.io/whitepapers/platforms/`

2
Enterprise Technology Landscape Overview

Before we start discussing the enterprise technology landscape, it is important to define what an enterprise is as it could mean different things to different people. An enterprise is a business or organization that is engaged in commercial, industrial, or service segments. Enterprises come in different shapes and sizes, from five-member startups to large multinational corporations with hundreds of thousands of employees distributed globally. They also operate in a variety of industries, such as manufacturing, retail, finance, technology, and healthcare. The main goal of a typical enterprise is to generate profits (apart from non-profits). Most enterprises also have other auxiliary goals, such as growth, innovation, social responsibility, and environmental impact. There are many different types of enterprises and corporations. Each type has its own legal and financial structure, and the type of enterprise that a business chooses will depend on a variety of factors, including the size and scope of the business, the industry it operates in, and the goals and objectives of the business. In summary, from modern cloud start-ups that were born in the last decade and have grown to be massive enterprises to centuries-old companies, enterprises cover a wide spectrum.

Getting back to the goals of this chapter, by the end, you will understand the application classification based on Gartner's pace layered architecture, the key characteristics of these layers, the corresponding infrastructure requirements, and how this diversity adds to the complexity in operating a hybrid cloud environment and how edge computing and **artificial intelligence** (**AI**) fits into the overall picture.

In short, we will have answered the following questions:

- How are applications and their supporting infrastructure classified under Gartner's pace layered architecture?
- How does this diversity make it difficult to create a standardized and efficient operating model?
- What impact does edge computing have on the already diverse enterprise IT environment?

The following topics will be covered in this chapter:

- Categorizing the enterprise technology landscape
- The diversity and complexity involved
- Difficulties in adopting a standard operating model

Categorizing the enterprise technology landscape

To better understand and manage the IT assets in the enterprise, a well-defined classification is required. There are many different ways to categorize applications; however, most organizations across industries follow Gartner's pace layered architecture to categorize their applications. We will be following Gartner's pace layered architecture as our application categorization framework in this book as well. A typical enterprise's application landscape consists of three different types of applications, as represented by Gartner's pace layered architecture approach, as shown here:

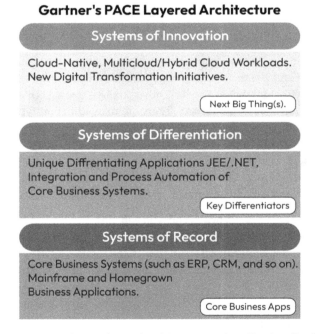

Figure 2.1 – Gartner's pace layered architecture and application distribution

As you can see, along with innovative next-generation applications under Systems of Innovation, there are a lot more applications under the **Systems of Record** (**core business apps**) and **Systems of Differentiation** (**unique capabilities**) layers that are key to operating the business. What we have observed is that for established enterprises, the total number of applications that are required to run the day-to-day business under **Systems of Record (SoR)** and **Systems of Differentiation (SoD)** are far greater than the applications under **Systems of Innovation (SoI)**. It begs the question: Should enterprises put all their digital transformation and cloud strategy focus on Systems of Innovation alone?

In practice, business value creation happens in three distinct phases:

- Innovate
- Operate
- Retire

Innovate is the key aspect of this cycle and might be the most chaotic as well, where faster provisioning, on-demand scaling, developer platform, market experimentation with a *fail fast, fail safe* approach, and so on are essential for success. The main aim of this phase is to prove a hypothesis and see whether there is a market or uptake of new differentiating features. This phase can significantly benefit from the public cloud environments where on-demand provisioning/scaling, self-service, and a *pay-as-you-go* approach are the norms.

However, once you establish the value and move to the *Operate* phase, you are tasked with different challenges. In this phase, standardizing the operations, availability, and scalability of the application platform to support business SLAs, security, and compliance, along with the ability to grow the business to new markets, becomes imperative.

The third phase, known as *Retire*, is about retiring applications with better replacements in place without any significant disruption and reallocating the resources for the next initiative. It can sometimes be challenging to find the right time to retire applications. Letting them live for too long will incur additional technical debt, which makes it more and more expensive and risky to maintain them. On the other hand, retiring them early will diminish their overall value. One approach to consider is scheduling periodic application retirement cycles to identify and retire applications in a standardized and streamlined manner. This will require a budget, planning, and a clear strategy. We will elaborate on this more later in this book.

In summary, enterprises need to create an IT environment where they can innovate faster, operate at scale, and retire services gracefully without compromising security and compliance at any time during this cycle.

It is also important to note that there is an industry-wide focus to migrate/modernize enterprise applications across these layers to be cloud-ready. That is, enterprise applications need to follow best practices such as cloud-native application architecture to maximize the benefits provided by the cloud, along with containerization and standard approaches such as rehost, refactor, and replatform, with varying degrees of success:

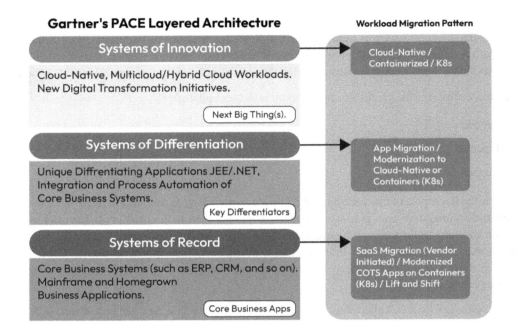

Figure 2.2 – Application migration pattern

As shown in the preceding figure, core business applications such as ERP and CRM systems are being actively moved to the SaaS model by their respective vendors. On the other hand, any **commercial off-the-shelf** (**COTS**) applications that need to be run in on-premises data centers are containerized. Some examples to highlight are Temenos and Amdocs. Temenos modernized its digital banking platform as containers to support on-premise, public cloud, and hybrid cloud deployment models (`https://www.temenos.com/news/2022/01/11/temenos-extends-strategic-collaboration-with-red-hat-to-deliver-digital-banking-automation/`). Amdocs built key new OSS/BSS applications as microservices running on Kubernetes containers to deliver new features and services to the market faster, as well as transform their entire development culture to focus on open, efficient work (`https://www.redhat.com/en/resources/amdocs-case-study`). Similar trends can be observed for applications in the Systems of Differentiation layer, where organizations are working on modernizing or migrating them to containers for better scalability and ease of operations. As far as Systems of Innovation applications are concerned, they are mostly born in cloud-native environments, and the few applications that were born before the cloud-native era are also being refactored at a faster pace (`https://www.konveyor.io/modernization-report/`). With **Kubernetes** (**K8s**) established as the container orchestrator of choice, enterprises can benefit significantly by leveraging a single consolidated platform that is built on enterprise-grade K8s to support their future applications and third-party services across these layers. We will discuss this in more detail later in this book.

The diversity and complexity involved

The enterprise landscape involves different types of applications with variable characteristics across the Systems of Innovation, Systems of Differentiation, and Systems of Record layers, as categorized in Gartner's pace layered architecture. These variations can be classified under two categories:

- Application architecture
- Infrastructure architecture

Let's take a look.

Application architecture

The application architecture differs from monolithic applications in Systems of Record to the cloud-native microservices architecture and functions in the Systems of Innovation layers. We'll look at the application architecture's characteristics and goals across these layers in this section.

Systems of Record

Systems of Record acts as the single source of truth for an organization's core business data and processes. Applications in the Systems of Record layer are typically built with traditional/legacy technologies, from mainframe to client-server to three-tier web application architectures leveraging established application frameworks such as JEE/.NET. The applications are typically as follows:

- Built as single deployment units, also known as monoliths
- Long-living, with infrequent upgrade cycles
- Follow the waterfall software development methodology, with longer development cycles and huge upfront investments
- Require significant downtime for upgrades
- Involve manual user acceptance testing, security and compliance reviews, and approval processes to ensure the business impact and risks are reduced
- Most probably purchased as COTS or built by SIs with varying contributions from internal IT teams

As a side note, some of these applications are being offered as **Software-as-a-Service** (**SaaS**) services as well. Given that SaaS applications are owned by the respective ISVs and organizations and consumers have little to no involvement in how these applications are built and managed, we will keep SaaS applications out of the scope and focus on the legacy Systems of Record applications that are still prevalent in most enterprises for the application characteristics and summary discussions provided.

The key characteristics of Systems of Record applications are as follows:

- **Not flexible**: Since they're built as monolithic applications with legacy technologies, these applications are not very flexible. Any change, be it big or small, will trigger a massive redeployment exercise, which can be time-consuming and costly.

- **Lesser deployment frequency**: Given the nonflexible nature of the application and the cost, risks, and time constraints associated with every deployment, Systems of Record applications are not deployed frequently. It is not uncommon for organizations to delay new deployments for as long as possible, stretching to multiple years in some cases.

- **Longer deployment downtime**: Systems of Record applications require significant application downtime during deployments. Since these applications are massive and typically require changes across the infrastructure, the storage and application server configurations, which are often done manually and verified, result in the rollout window being in hours if not days. Hence, the deployment is scheduled during weekends or public holidays to avoid any business disruption and to ensure all security and compliance checks are done as part of the deployment. Most IT contributors might still be involved in dreaded year-end deployments and war room discussions that come with this.

- **Significant business impact**. Given the critical nature of Systems of Record applications in delivering core business functions, any downtime that's incurred will have a significant business impact. Hence, every deployment is carefully analyzed and orchestrated to improve the chances of a successful rollout.

- **Outdated security architecture**: Most Systems of Record applications were built to be deployed inside corporate data centers and accessed by internal employees. Given this, they can be susceptible to some common security threats, such as insider threats, compliance violations, malware and ransomware attacks, and data breaches.

In summary, Systems of Record applications are historically heavy, less flexible, require deployment downtime, and when not available, cause significant business disruption.

Systems of Differentiation

Applications in the Systems of Differentiation layer help organizations differentiate themselves from their competitors by providing unique capabilities or services. Typically, these applications are built and maintained internally, given the unique differentiating capabilities they serve for their business. These applications can also be COTS or SaaS applications but are heavily customized to implement unique differential capabilities that organizations provide. Systems of Differentiation applications integrate and extend Systems of Record applications and data, so they are typically impacted by the changes that happen in Systems of Record systems. The applications in the Systems of Differentiation layer share more characteristics with Systems of Record applications, such as technologies used and the release management and life cycle management processes that are followed. Additionally, they leverage integration, extension, and process orchestration technologies such as service bus, **service-oriented** architecture (**SOA**), BPEL/BPMN, and others to access Systems of Record applications and data systems. One note to highlight here is that traditional integration, extension, and process orchestration implementations are modernized to a more cloud-native architecture, leveraging microservices and APIs to make them more flexible, open to change, and loosely coupled.

The key characteristics of Systems of Differentiation applications are as follows:

- **Less flexible**: Built with a similar architecture as Systems of Record applications, Systems of Differentiation applications are typically rigid and built as monolithic applications with legacy technologies. However, the integration, extension, and process orchestration technologies are relatively more flexible and can evolve independently. Hence, overall, Systems of Differentiation applications are a little more flexible than Systems of Record applications.

- **Higher deployment frequency**: Systems of Differentiation applications are deployed more frequently than Systems of Record applications since they have a smaller business impact and are also integrated and orchestrated better with standardized and loosely coupled architectures that leverage service bus, SOA, BPEL/BPMN, and others.

- **Lesser deployment downtime**: Depending on what you are deploying, the deployment downtime for Systems of Differentiation can vary. For monolith applications, it can be similar to if not the same as Systems of Record applications, and there will be significantly less downtime for integration, extension, and orchestration changes. So, depending on the change and the application type, Systems of Differentiation rollouts can be scheduled during business hours or follow the same model as Systems of Record applications.

- **Marginal business impact**: Given that Systems of Differentiation applications typically do not deliver core business functions, the downtime that's incurred is less impactful than it is for Systems of Record applications. Hence, depending on the organization's internal release processes, Systems of Differentiation applications can be deployed more frequently with little risk and downtime compared to Systems of Record applications.

In summary, Systems of Differentiation applications can be heavy, but they are more flexible, require reduced deployment downtime, and when not available cause marginal business disruption compared to Systems of Record applications. While Systems of Record applications focus on maintaining the accuracy and integrity of data, Systems of Differentiation applications focus on leveraging the data and technology provided to create a competitive advantage for the organization.

Systems of Innovation

The Systems of Innovation tier was originally built as the enterprise innovation factory, where new ideas were built and put to the test, and dismantled after their purpose was served. Once proven, winning applications were typically rebuilt as either Systems of Differentiation or Systems of Record applications based on the purpose they served and the organization's application classification. Systems of Innovation applications are built with modern techniques such as polyglot development, microservice functions, and so on, and leverage as much automation as possible to ensure fast development and delivery cycles. Most of the techniques that are built in this layer eventually become mainstream and are eventually adopted as a common development and delivery mechanism across the enterprise. A good analogy for this is **Formula 1 (F1)** racing, where competing teams innovate at speed to get an advantage over the other teams. After a few cycles of testing and perfection, these innovations become mainstream and are introduced to passenger cars (`https://f1chronicle.com/how-advanced-technology-gets-transferred-from-formula-1-to-production-cars/#:~:text=Enhanced%20Efficiency%20and%20Battery%20Technology,-Formula%20One%20technology&text=Lithium%2Dion%20batteries%20are%20rechargeable,improving%20their%20fuel%20consumption%20patterns`).

Systems of Innovation applications are typically as follows:

- Built as a collection of microservices and serverless functions (**Functions-as-a-Service (FaaS)**)
- Short-lived with frequent automated deployments of individual microservices with minimal/no application downtime
- Follow Agile principles and do not incur significant upfront investments.
- Practice/strive for zero downtime upgrades
- Leverage automated testing as much as possible, with inbuilt security and compliance standards (when required)
- Built by internal IT working closely with customer-facing business units and rarely outsourced

The key characteristics of Systems of Innovation applications are as follows:

- **Extremely flexible**: Built with a modern application architecture that leverages microservices and functions, Systems of Innovation applications are extremely flexible because the microservices can evolve independently and push the latest code to production with minimal application downtime.
- **Maximum deployment frequency**: Given the modern application architecture and development practices and automated deployment model with in-built security and compliance, Systems of Innovation applications can be deployed often on demand, thus aiding a faster innovation cycle.
- **Minimal/no deployment downtime**: With automated and canary/blue-green deployment models, Systems of Innovation applications strive to achieve zero downtime and in most cases wish to push changes to live production environments during normal business hours without any downtime.
- **Minimal/no business impact**: Given the experimentation aspect of Systems of Innovation, they typically do not have a direct business impact. In some cases, when the applications support live customers during the validation phase, minimal business impact can occur. However, with automated and zero downtime deployment models, those issues are addressed swiftly.

In summary, Systems of Innovation applications are lightweight thanks to the modern development and delivery model, extremely flexible, deployed on demand, and when not available cause no significant business disruption.

The following figure highlights the characteristics of these layers:

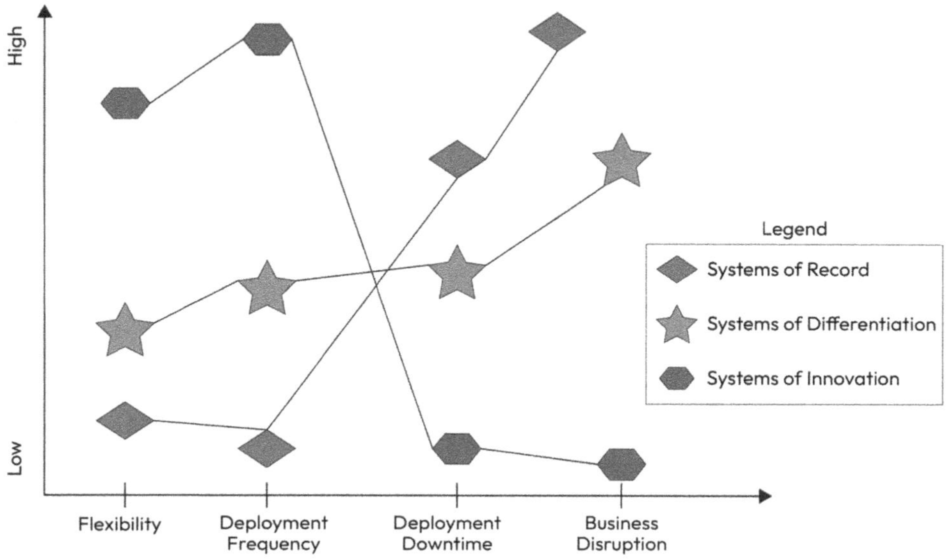

Figure 2.3 – Gartner's pace layered architecture – application characteristics

In the next section let us explore the characteristics of the infrastructure across these three layers.

Infrastructure architecture

Infrastructure refers to the underlying technology and resources that support your applications and workloads. Infrastructure includes hardware, software, networking equipments, data centers, servers, and storage devices. Similar to application architecture classification, infrastructure can also be classified into the same three layers because the requirements and the behavior of the infrastructure differ from layer to layer – that is, the infrastructure requirements for the Systems of Record layer will be different from that of the Systems of Differentiation layer, which, in turn, will be different from that of the Systems of Innovation layer.

Systems of Record infrastructure

Systems of Record applications play a critical role in delivering core business applications for the enterprise. Given this, these applications must be highly reliable and stable. They must also have a well-thought-out and tested scalability and disaster recovery architecture in place. To support these needs, it is important to use an infrastructure that is highly reliable, scalable, and secure. This may involve using enterprise-grade hardware, such as servers, storage systems, and networking equipment that is often vertically integrated and supported by a single vendor.

Systems of Record applications may also require specialized infrastructure to support specific requirements, such as high-performance computing or data analytics. To address these requirements, most organizations build their Systems of Record infrastructure on-premises with a well-engineered, powerful, and often expensive hardware ecosystem, from mainframe to virtualized server farms built with hyper-converged infrastructure. Organizations also invest in state-of-the-art encryption, backup and restore, and disaster recovery infrastructure to ensure the Systems of Record applications are as available, resilient, and secure as possible.

Systems of Differentiation infrastructure

The Systems of Differentiation layer consists of applications that help organizations provide differentiated capabilities/services and integration, extension, and process automation capabilities that help these systems talk to the Systems of Record applications and data sources. Given this, the Systems of Differentiation infrastructure layer may need to be as resilient and secure as the Systems of Record infrastructure and also more scalable, especially for the integration, extension, and process automation infrastructure. However, the Systems of Differentiation infrastructure typically may not include legacy, vertically integrated servers such as mainframes, which are used to deploy core business applications and rely on virtualized infrastructure built on commodity or hyper-converged infrastructure. In summary, the infrastructure requirements for Systems of Differentiation depend on the specific needs and goals of the organization and the nature of the applications being supported. In general, they are as resilient and secure as Systems of Record applications and more scalable to ensure that differentiated capabilities are leveraged better during peak seasons to improve business output. Organizations also leverage public cloud infrastructure more aggressively to build some of these differentiating capabilities with cloud-native app development practices.

Systems of Innovation infrastructure

Systems of Innovation applications are typically short-lived, distributed, and horizontally scalable, and leverage modern cloud-native development and delivery practices. Given this, the Systems of Innovation infrastructure needs to be as follows:

- Provisioned on-demand or via self-service
- Highly available
- Elastic
- Horizontally scalable
- Loosely coupled
- Have automated patching and upgrade support

The adoption of the public cloud infrastructure is significant for Systems of Innovation since the public cloud inherently supports all these requirements out of the box. As more organizations adopt digital transformation to embrace a digital-only/digital-first business model, the Systems of Innovation infrastructure is fast becoming the enterprise infrastructure standard to help organizations focus on innovation velocity to deliver business outcomes and be able to operate at scale with inbuilt security and compliance.

The following diagram shows the various infrastructure characteristics across the three layers:

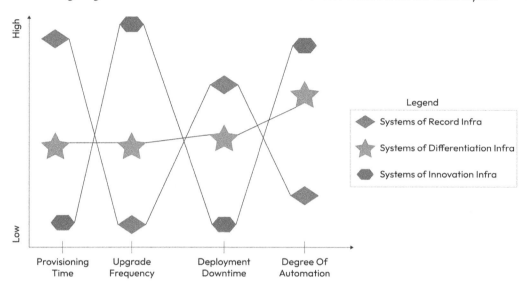

Figure 2.4 – Gartner's pace layered architecture – infrastructure characteristics

Depending on the hardware used, the Systems of Record and Systems of Differentiation infrastructures can be more automated to deliver faster provisioning and shorter deployment downtime.

In summary, organizations need different types of infrastructure, from high-cost, vertically integrated, and long-living infrastructure to low-cost, commodity, horizontally grouped, short-living, and nimble infrastructure to support the three different application categories explained here. From a deployment perspective, it may span from edge devices to the core data center, to the cloud. This diversity in infrastructure and deployment models makes it difficult for organizations to build and manage a consistent deployment model.

Difficulties in adopting a standard operating model

As explained earlier in this chapter, organizations have a diverse application and infrastructure landscape with unique characteristics based on their architecture and the business requirements they serve. Therefore, it is difficult to adopt a standardized operating model across the layers since emerging technologies such as edge computing and AI will further increase this complexity. Focusing on just the infrastructure layer alone, the Systems of Innovation layer delivers the most value with faster and safer experimentation capabilities to test new ideas, end-to-end automation, self-service, and horizontal scalability. But for enterprises that deliver most of their business services with traditional Systems of Record applications, moving to a modern infrastructure may not be feasible. Two opposing viewpoints are in play:

- Resist change to reduce risk and improve stability
- Embrace change to improve innovation velocity and reduce the risk of irrelevance

At this point, it is worth mentioning the variables that edge and AI deployments bring to the mix. Edge deployments cover both edge data centers and edge devices. Irrespective of where they are deployed, the applications and their infrastructure exhibit certain characteristics that are unique and quite different from data center and cloud environments. Some of the key differences are listed here:

- **Disconnected**: Often, edge deployments are deployed far out from the main data center/cloud over an unreliable network connection. Given this, the edge deployments are built to operate independently. The upgrade cycles are also typically managed differently for edge deployments given this limitation.
- **Limited capacity**: Most edge deployments are constrained with limited capacity infrastructure. This is done to simplify the overall life cycle management of the infrastructure and also due to limited skillsets/shortage of IT resources in the edge locations. This could add additional types of hardware for edge deployments that are not otherwise used in on-premise data centers and cloud infrastructures.
- **Special toolset**: Edge infrastructure and applications may not be compatible with the on-premise and cloud deployments and hence will require different tool sets and processes for application development, delivery, security and compliance, and ongoing maintenance.

Apart from these aspects, based on the use case, the number of devices that are part of edge deployments can be magnitudes higher than the number of servers in on-premise and cloud deployments. Device management is a unique and complicated process that more organizations are starting to figure out as they build and deliver more edge capabilities.

Similarly, AI deployments bring their own unique footprints, some of which are highlighted here:

- **Specialized hardware**: To handle complex computations and calculations quickly and efficiently, AI systems often use specialized hardware such as GPUs or TPUs, which are different from the typical hardware that's used in on-premise DCs and cloud environments.

- **New technologies**: AI also requires development tools that are more open source aligned and are typically unique and not widely used in organizations.

- **Storage and security**: Another key aspect of AI systems is the requirement to support enormous amounts of data to help train the systems on an ongoing basis for better efficiency and accuracy. This requires organizations to leverage different storage tiers to reduce costs, along with implementing better data security.

Resist change to reduce risk and improve stability

Traditionally, organizations build operating models that focus on resisting changes to improve the availability and stability of their applications. The more important the applications are, the greater the resistance. This is evident in the way the core business applications that are categorized under Systems of Record are built and managed. Here are some key aspects to focus on:

- **Waterfall development model**: The waterfall development model is a sequential software development process where various phases of the software development life cycle are executed in sequence without any overlap. The phases of the waterfall model include requirements gathering, design, implementation, testing, deployment, and maintenance. This model is often used for projects with well-defined requirements, a clear end goal, and a predefined scope. It is considered to be a traditional method of software development that fits well for developing core business applications where the requirements are documented end to end, the project is scoped and budgeted in advance, and long development, testing, and acceptance timelines are defined.

- **Tightly coupled monolithic applications architecture**: The application architecture of most of the core business/ISV applications follows a traditional two-tier or three-tier application architecture. These applications are built as a single deployment unit due to the tightly coupled nature of the architecture and also to reduce deployment complexity.

- **Manual/less automated release management**: Given the longer duration of application development and the delivery process adopted by the waterfall development model, there is ample room for manual and sequential release management to ensure that all security and compliance requirements are addressed before the applications are deployed for use. Depending on the application's type and the organization's IT maturity, certain aspects of the release management process may be automated for better efficiency and control. However, the purpose of the automation is more centered around consistency and error mitigation and less on development velocity.

- **Infrequent upgrade cycles**: The sequential nature of the release process and the potential business impact the upgrade cycles can cause due to application downtime makes it difficult for organizations to deploy changes to the production environment frequently. Typically, organizations combine multiple requirements under one release and plan on two or three releases per year. These deployments are planned during public holidays or weekends to minimize business disruption. I am sure that most of us are familiar with the "*Website is down* *for maintenance, please try again at 9:00 A.M. on Monday*" message that used to be displayed during maintenance/upgrades from a few years back. One interesting question to ask is, "*When did you see this message last?*" We believe this was probably quite a while back for consumer web and mobile applications as more organizations moved away from this approach toward a more agile development model.

- **Siloed operations**: From an infrastructure operations perspective, we typically see organizations adopt a more siloed approach, with dedicated teams with clearly defined SLAs for the services they offer. The **Information Technology Infrastructure Library (ITIL)** is predominantly used for managing IT services.

In summary, the goal is to resist change with predefined and structured development and delivery processes to reduce risk and increase stability. Every change is considered to be risky and hence they are delayed and combined to avoid frequent releases. The focus is on following the established process with as little deviation as possible, with various checkpoints along the way. The outcomes that are achieved are not the main focus.

In the modern era, where applications are expected to always be available and also evolve to meet ever-growing customer and business expectations, this approach may not be a good fit. Organizations have modernized away from this model with varying degrees of success. One such story is from Best Buy, the world's largest electronic retailer. Joel Crabb, one of the architects in the digital transformation team, explains the difficulties in making small changes to the monolithic e-commerce portal. He explains how simply widening the product details page and moving the shopping cart button from left to right cost Best Buy a million dollars and took 6 months to implement. The key reasons he attributes to this complexity are as follows:

- Monolithic application architecture
- Shared development environment with a backend database dependency
- Manual regression testing
- Siloed org structure

He has created a 22-part series documenting this journey at `BestBuy.com` (`https://joelcrabb.com/?p=334`).

Embrace change to reduce risk and improve innovation velocity

Modern Systems of Innovation applications are built with cloud-native architecture and automated application delivery processes to improve innovation velocity. They are also built with automated and elastic cloud infrastructure. The overall focus is to deliver faster business outcomes with increased application development and deployment velocity and efficient operations. Here are some of the key characteristics:

- **Microservices architecture**: Applications are built as loosely coupled independent services that can be built, updated, and deployed independently. All communication between the services is done through clearly defined APIs.

- **Agile development methodology**: Agile development is practiced to improve collaboration, flexibility, and faster innovation. Agile is characterized by iterative, incremental development, where requirements and solutions evolve through the collaborative effort of self-organizing and cross-functional teams. Some of the most popular Agile development methodologies that are followed are Scrum, Kanban, and Lean development.

- **Automated application delivery**: **Continuous integration** and **continuous delivery/deployment (CI/CD)** pipelines are leveraged to help development teams build, test, and release software more frequently and quickly. CI/CD pipelines cover key delivery tasks, such as building and testing code, running automated tests, packaging and deploying code, and monitoring the deployed software. With this, teams can reduce the time and effort required to release new features and can deploy new features and bug fixes more frequently at speed.

- **Cloud infrastructure**: To achieve these outcomes, organizations practice cloud deployment architecture. Cloud infrastructure offers self-service for developer provisioning and automated life cycle management for highly available and secured infrastructure layers to support modern Systems of Innovation applications.

The core focus is to be able to build and deliver software at speed with automated security and compliance. The focus is aligned with outcomes and the processes are automated with reports and metrics created on demand. The goal is to achieve maximum efficiency across development and operations to leverage technology to deliver maximum business outcomes, from the ability to release new features to staying ahead of the competition, piloting new ideas to create new revenue streams, and improving the customer experience.

Many organizations, especially in industries that are heavily disrupted by modern cloud start-ups such as retail and FSI, have leveraged this approach in the past decade. Just to stick with our retail example, one such success story comes from Home Depot, the largest home improvement retailer in the world. The saying goes that Home Depot decided to digitally transform in 2017 after finding out that a popular online retailer sold more hammers than them. With this renowned focus and execution, Home Depot was able to successfully address the challenges posed by COVID-19-induced disruptions. According to their 2020 annual report, digital sales grew 86% versus the year prior, their digital properties had record traffic throughout the year, and overall sales increased by 20%.

As you can see, depending on the application architecture and the supporting infrastructure characteristics, applications are built and operated to resist change or embrace change. More and more organizations are building and delivering modern applications and infrastructure to be able to embrace change. Given the benefits it provides, legacy applications and infrastructure that traditionally resisted changes are also being migrated and modernized to modern architecture to embrace change. This journey might take a few years to complete and in the meantime, most organizations will end up with a mix of both applications and infrastructure types. This mix makes it difficult for organizations to build and manage one operating model for all their applications and infrastructure. Most organizations will start with at least two operating models (if not more) – one to support the applications and infrastructure that resist change and another for applications and infrastructure that embrace change. Putting this back in the context of Gartner's pace layered architecture, there is one operating model for Systems of Record and most Systems of Differentiation applications and infrastructure and a separate one for Systems of Innovation and a few Systems of Differentiation applications and infrastructure that are closely aligned with the Systems of Innovation architecture and deployment model.

Summary

In this chapter, we explained how Gartner's pace layered architecture classifies enterprise applications across the Systems of Record, Systems of Differentiation, and Systems of Innovation layers. We also walked through how the application characteristics differ across these layers and how they impact the infrastructure layer that supports them. This results in a diverse environment with different priorities and change resistance, which makes it difficult to adopt a standardized operating model across the organization.

In the next chapter, we will explore how enterprise IT departments adopted a bimodal approach to address these conflicting priorities, and the resulting learnings to be taken into consideration as we adopt new emerging technologies.

Further reading

The following links provide more information about the topics covered in this chapter and allow you to further deep dive into any topics of interest:

- Accelerating Innovation by Adopting a Pace-Layered Application Strategy: `https://www.gartner.com/en/documents/1890915`

- What is IT Infrastructure Library (ITIL)?: `https://www.ibm.com/sg-en/topics/it-infrastructure-library`

3

Learnings From Bimodal IT's Failure

As we discussed in the previous chapter, most organizations have a diverse application landscape with applications distributed across Systems of Record, Systems of Differentiation, and Systems of Innovation, all of which require different types of infrastructure and operating models to support them. To address this, organizations typically have multiple different approaches to managing core business applications and the supporting infrastructure with one model and the more innovative, fast-moving, experiment-focused applications and the supporting infrastructure in another model. Gartner came up with a no-longer popular model for this called **bimodal IT**, which talks about how organizations can build and manage two different IT operation modes to support the diversified applications and the corresponding infrastructure landscape. As part of this chapter, we will introduce the concepts behind bimodal IT, how it was operationalized, the limitations of the model, and why it is not progressive and flexible enough to make organizations succeed in the digital and cloud era that we are focusing on in this book. By the end of this chapter, we will have answered the following questions:

- What is bimodal IT?

- How have organizations adopted bimodal IT to manage the diversified IT environment?

- Why is bimodal IT not good enough for the distributed future that most organizations are moving toward?

The following topics will be covered in this chapter:

- Introduction to bimodal IT

- Enterprise IT management and bimodal IT

- The challenges and limitations of bimodal IT in the distributed future

Introduction to bimodal IT

Gartner introduced bimodal IT in 2014 as a way for organizations to better manage their IT operations and stay competitive in the digital era. Gartner defines bimodal IT as *"the practice of managing two separate, coherent modes of IT delivery, one focused on stability (Mode 1) and the other on agility (Mode 2)."* Mode 1 focuses on maintaining and managing the existing technology landscape, while Mode 2 focuses on quickly developing and deploying new applications/features to meet the changing needs of the business. The goal of bimodal IT is to align IT with the overall goals of the organization and simplify complexity and manage costs while maintaining the necessary level of service. Mode 1 is traditional and sequential, with a focus on safety and accuracy. From the previous chapter, where we covered applications and infrastructure categorization, you can think of Mode 1 as the model that supports Systems of Record applications and traditional Systems of Differentiation applications and infrastructure, that is, it provides a structured, sequential, and stable IT operations model for systems that resists change and their associated infrastructure. The highest priority is given to stability, availability, and security in this mode. On the other hand, Mode 2 is more agile and exploratory; it emphasizes agility and speed by providing a more fluid environment for the rapid development and deployment of new ideas, where Agile methods and a DevSecOps provisioning model allow for greater business involvement and fast iterations. It maps to Systems of Innovation and a few modern Systems of Differentiation applications and infrastructure. The following are the proposed outcomes of bimodal IT:

- **Improved agility**: By separating IT operations into two distinct modes, organizations can be more agile and responsive to changing business needs

- **Better risk management**: Mode 1 provides a stable and secure environment for critical systems, while Mode 2 allows for more risk-taking and experimentation

- **Greater innovation**: Mode 2 allows for the iterative development of new digital products and services, which can help organizations stay competitive in rapidly evolving markets

- **Increased collaboration**: Bimodal IT can help break down silos between different departments and encourage more collaboration and communication between IT and other business units

- **Improved customer experience**: By leveraging Mode 2 to develop innovative digital solutions, organizations can deliver better customer experiences and meet changing customer needs

Another way to look at this is in terms of two-speed IT (the practice of managing two separate modes of IT delivery, one focused on stability and the other on agility), Mode 1 is "slow or low change cadence IT" for Systems of Record and traditional Systems of Differentiation, while Mode 2 is "fast or high change cadence IT" for Systems of Innovation and modern Systems of Differentiation. The strengths of Mode 1 development teams are often weaknesses in Mode 2 development and vice versa. There is a general agreement that some development projects require a more traditional approach, while others benefit from Agile methods. It is critical to note that, according to Gartner, bimodal IT didn't necessarily mean having two separate teams, but rather having two ways of working, with different processes, tools, and metrics. It allows organizations to have both a stable, predictable IT environment for critical systems and an agile, flexible environment for new initiatives, while also separating IT operations into two modes, each with its own set of processes and tools. The following table shows how these two modes can be compared across different factors:

	Mode 1	Mode 1
Goal	Reliability	Agility
Value	Price and Performance	Revenue, Brand, customer Experience
Approach	Waterfall	Agile
Process	Manual, Sequential	Automated, Parallel
Culture	IT Centric	Business Centric
Cycle Time	Long (Months)	Short (Days, Weeks)

Table 3.1 – Bimodal IT comparison

However, most organizations that practice bimodal IT have taken this a step further by separating Mode 1 and Mode 2 into different groups within an organization. In addition to their differences in development methods and tools, the Mode 1 and Mode 2 development groups don't report through the same organizational structure and don't use the same delivery mechanisms. This is done to reduce conflicts for resources and budget within IT, which is currently tasked both with stable current operations and innovative future results. We can think of Mode 2 capabilities being represented by shadow IT and citizen developers embedded within business units. This has become more and more common in the digital transformation era. Bridging Mode 1 and Mode 2 is a critical challenge most organizations are tasked with. Based on our discussion later in this chapter, we will see why it is not a simple task that often leads to overall complexity and failed adoption.

Enterprise IT management with bimodal IT

Thanks to the demands of employees and customers on enterprise IT under digital transformation, IT must be faster and smarter in the face of digital disruption, rapidly evolving markets, and new ways of working and doing business while still retaining the control and oversight required to guarantee compliance and security. Bimodal IT addresses this challenge by balancing the stability, safety, and accuracy of legacy in-house IT investments (also known as Mode 1) with the agility, speed, and innovation of continuous delivery through cloud services and applications (also known as Mode 2). Gartner predicted that organizations that can master bimodal IT will be in a strong position to compete in the digital business era, while those that cannot will struggle to keep pace with the rapidly changing IT landscape. Some examples of bimodal IT exploration are as follows:

- ING Bank
- General Electric
- Telstra

Most of these implementations have common patterns in which the organizations leveraged the same pattern, with Mode 1 primarily focused on Systems of Record and Mode 2 on Systems of Innovation. Here are some ways in which organizations have operated bimodal IT:

- **Dedicated IT infrastructure and tools**: To support the different needs of each mode, it may be necessary to create separate IT infrastructure and tools. For example, traditional IT operations may require a more robust and stable infrastructure, while agile innovation projects may require a more flexible and scalable infrastructure. Most organizations follow this model to create a seamless environment for both Mode 1 and Mode 2.

- **Defined roles and responsibilities**: In bimodal IT, it is critical to clearly define the roles and responsibilities for each mode of IT delivery. This includes identifying the teams and individuals responsible for each mode, as well as the processes and tools that will be used. It may very well require creating new roles for Mode 2 that are more aligned with modern development and delivery mechanisms, such as DevOps engineers. DevOps engineers reduce that complexity, closing the gap between actions needed to quickly change an application and the tasks that maintain its reliability. Roles such as DevOps engineers typically do not exist in Mode 1.

- **Established governance and decision-making processes**: Mode 1 and Mode 2 require different governance and decision-making processes and hence should be built separately and tailored to the specific needs of that mode. Typically, Mode 2 requires more automated operations, while Mode 1 implements more sequential and manual operations. In other words, Mode 1 operations require a more formal, centralized decision-making process, while Mode 2 operations thrive on a more decentralized, self-organizing approach.

- **Enhanced communication and collaboration**: Communication and collaboration between the different teams and individuals involved in bimodal IT is essential. Inside the models, Mode 2 leverages modern communication platforms such as Slack, Google Workspace (earlier known as gSuite), and others for enhanced collaboration. On the other hand, Mode 1 may still leverage traditional communication tools such as email, chat, and ticketing systems. Across the teams, improved collaboration is achieved through information sharing, collaboration, cross-functional teams, and other mechanisms.

- **Measured, evaluated, and improved**: One of the key aspects of Mode 2 is to continuously measure the efficiency of operations and evaluate and improve them on an ongoing basis. This approach can be extended to the overall bimodal IT to ensure that the organization is achieving its goals for the bimodal IT implementation.

The following figure, which is based on Gartner's solutions variance, shows the differences between these modes, wherein Mode 1 is represented as *Traditional Mode* and Mode 2 is represented as *Nonlinear Mode*:

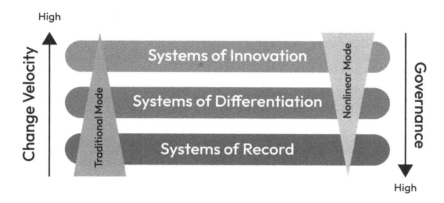

Figure 3.2 – Bimodal solution variants

The key takeaway is that most organizations operate bimodal IT with two different teams with varying roles and responsibilities, tools, and processes. Even though this may yield the desired results in helping organizations address the traditional and modern needs they are tasked with, it fails to address some of the key areas that organizations face now as they embark on digital transformation and cloud adoption to propel them to the digital-first era that we are in. In the next section, we will explain why bimodal IT fell out of favor and is mostly classified as a failure.

The challenges and limitations of bimodal IT in the distributed future

With digital transformation and cloud adoption accelerating in the last few years, all the corresponding changes that IT organizations are implementing to support this has added new unforeseen complexities that bimodal IT is not built to handle. One aspect that was highlighted by many organizations was that bimodal IT is not new. Irrespective of Gartner creating a model and guidelines, many organizations implemented bimodal IT based on the model they were already familiar with by building two separate teams and operations model, which created two incompatible and often rivaling visions and priorities that organizations were trying to get away from in the first place. As early as 2017, **International Data Corporation (IDC)**, in their 2017 CIO outlook, predicted that by 2019, the CIOs who had implemented bimodal IT will accumulate a crippling technical debt, resulting in spiraling complexity, costs, and lost credibility. Forrester also had similar sentiment in their 2017 article titled *Bimodal IT is past its due date* (`https://www.forrester.com/blogs/bimodal-it-is-past-its-due-date-providing-speed-and-innovation-need-to-top-the-cios-agenda/`, `https://www.forrester.com/report/The-False-Promise-Of-Bimodal-IT/RES131967`).

From a project management perspective, Mode 1 and Mode 2 bring in additional challenges as well. The research article titled *Management Challenges in Bimodal IT Organizations* from the Journal of

Information Systems Engineering and Management (`https://www.jisem-journal.com/download/management-challenges-in-bimodal-it-organizations-12014.pdf`) details these differences. They can be seen here:

Table 1. Mapping of Mode-1 and Mode-2 operations toward project management principles.		
PM principles	Mode-1	Mode-2
1. Work organization	SDLC/V-Model phases	Sprints (usually biweekly, time boxed development of ready for production features)
2. Estimations	Effort estimation based on the Work Break-down Structure for each phase of the SDLC/V-Model phases	High level estimation and then spring planning using story points, a measure of complexity of the work to be accomplished. Also capacity based estimation sometimes used, instead of story points.
3. Prioritization of work	According to the SDLC/V-Model phases	According to perceived business value of use cases: higher value first, then the lower ones
4. Deliverables	Interfaces, reports, processing blocks and application modules	Implementation of user stories (functional analysis) that comprise use cases
5. Time	Is an element of the time plan (Gantt chart)	Is a box that contains implementation delivery targets
6. Staffing	People participate into phases and steps according to their roles & job profile.	Dedicated people that participate from inception to completion of the project and play potentially several roles within the team.
7. Progress reporting	Actuals vs. baseline	Burn down chart
	The Earned Value Measurement System (EVMS) has become a mainstay in Commercial and Government groups to measure progress and success of a project. EVMS is espoused to be an effective (albeit subjective) measure, but it does not play well with agile development efforts, due to its requirement of static schedules and work plans. Here we introduce a new paradigm for EVMS that will accommodate and be effective in measuring progress and problems within agile development efforts." (Crowder and Friess, 2015). A similar investigation has been previously examined by(M. Griffiths and A. Cabri, 2006).	
8. Scope of work and peer deliverable	As defined in the SDLC/V-model phases.	Definition of "done"? per sprint.

9. Work validation	Based on alignment of the V-model's corresponding elements per phase.	Customer acceptance using Net Promoter Score (Grisaffe, 2007) or other marketing KPIs.
10. Learning	At the end of the project, usually documented at the project close-out report.	At the end of every sprint, a retrospective meeting calls for evaluation of success and failure points recognized by the team members themselves, for faster learning curves.
11. Documentation	Exhaustive descriptions, including data- and work-flows.	Minimal, however additional material such as high-level analysis trees/ideas configuration etc. are needed. Wiki repositories are sometimes used, with references to use cases (business requirements) and user stories (functional analysis specifications)
12. Project manager	As defined in PMBOK - safeguards deliverables delivery on time and within budget, with the available resources at the best possible quality.	"The agile manager understands the effects of the mutual interactions among a project's various parts and steers them in the direction of continuous learning and adaptation." (Augustine et al., 2005)
13. Tools	Traditional project management and software development tools.	Commercial agile tools (Mihalache, 2017)

Figure 3.3 – Management Challenges in Bimodal IT Organizations, by
Journal of Information Systems Engineering and Management

In summary, here are some of the key reasons why bimodal IT fell out of favor:

- **Poor governance**: Often, the main reason for bimodal IT failure is attributed to the lack of a clear governance model. This happens more often with organizations treating the two modes as totally independent entities. The governance structure that is put in place typically works well within their respective modes but fails when it is applied across the modes. There are no clear guidelines for implementing bimodal IT across the modes, and each organization may have a different approach. This can make it difficult for organizations to know how to implement it effectively.

- **Increased complexity**: Implementing bimodal IT can be complex as it requires creating two separate IT delivery modes and ensuring that they work together. This can be difficult to achieve in practice and may require significant changes to be made to existing processes and systems.

With organizations operating Mode 1 for a longer period, the changes and the collaboration required to make bimodal IT work may be difficult to embrace, resulting in operational complexity leading to a less collaborative and brittle environment. This issue can also be extended to the business stakeholders who may now have to work with two different IT teams operating on two different modes to get their work done. Some of this can be addressed with centralized IT functions such as enterprise architecture and PMO practices, but this may lead to additional bottlenecks, thus diminishing the overall efficiency.

- **Lack of ownership**: With IT efficiency becoming more and more important to deliver faster business outcomes to support the digital era, the concept of bimodal IT is seen as very limiting and rigid. To deliver the modern demands of digitalization, all three classifications of Systems of Innovation, Systems of Differentiation, and Systems of Record across Mode 1 and Mode 2 need to evolve at a faster pace. This fundamental change in which stability and innovation evolve together involves both Mode 1 and Mode 2 changing, along with the application architecture, data architecture, and infrastructure life cycle management. If organizations implement bimodal IT as two distinct models, then the collaboration that's needed to make the necessary changes, along with the required ownership, is not addressed, making it a difficult proposition.

- **Lack of architectural guidance**: While Gartner conceptually presented bimodal IT, it did not explain how systems should be architected across these two modes. This not only leaves lots of room for misinterpretation but also for costly architecture and design mistakes that may have a significant impact on delivering services to the customers.

The following diagram shows an example of a well-architected system that addresses this concern:

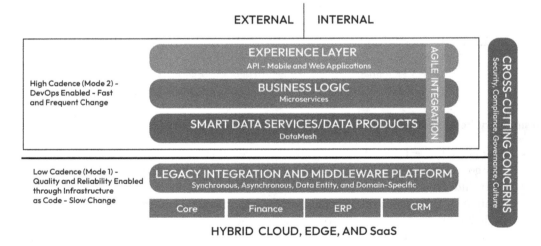

Figure 3.4 – A possible bimodal architecture

This architecture shows how the data services system architecture is represented across Mode 1 and Mode 2.

- **Distributed future**: One additional aspect to highlight here is the distributed future thanks to edge computing with fast-evolving 5G, 6G, and IoT adoption and the notable rise of AI and innovative applications that are built on it. These two trends add additional dimensions to the IT environment that Bimodal IT is not structured to handle. As explained in the previous section, the deployment model, the new infrastructure and tools leveraged, and the additional storage and security needs these new systems bring in may require adding a different mode – that is, Mode 3 – on top of Mode 1 and Mode 2 that most organizations currently operate. This further complicates the issues that we have highlighted here.

In summary, bimodal IT is not a novice concept for most organizations. Traditionally, organizations follow two different modes of operations – one for the core business applications, which are typically more locked in with sequential and structured release management practices, along with stringent security and compliance requirements, and another for more exploratory, innovative, and experimental applications. This trend has continued even after digital transformation and cloud-native development practices went mainstream, where Mode 2 was established as the preferred mode of operations. We still see certain legacy applications and infrastructure being categorized as too risky to change and are managed with Mode 1 operations. Irrespective of what we call it, either bimodal IT or Greenfield and Brownfield, the approach of managing two different practices seems to be more common than we think.

Another point to highlight here is the increased focus on application modernization. Unlike over the past three decades, recently, there is a significant interest and focus on application modernization. The main reason for this is that applications need to be altered to be cloud-native, cloud-optimized, or cloud-ready to be able to leverage the value that Mode 2 delivers via cloud computing and the corresponding development and delivery processes around it. Some of the key attributes for modern cloud-aligned Mode 2 are listed here:

- **Self-service**: Ability to provision environments/services on demand
- **Scalability**: Ability to allocate additional resources seamlessly to support any additional load
- **Elasticity**: Ability to dynamically leverage resources to grow and shrink capacity
- **Flexibility**: Ability to respond to new challenges and opportunities with technology
- **Accessibility**: Ability to access the systems via different channels, from web to mobile
- **Security**: Ability to respond to security threats quickly with minimal disruption
- **Reliability**: Ability to deliver a high-quality service consistently with minimal disruption
- **Multi-tenancy**: Ability to share the platform with different applications/services to improve overall efficiency and reduce operational costs

To leverage these benefits across the organization, organizations are modernizing their applications at a faster pace so that they can be operated under Mode 2. However, this requires a lot more than a simple technical process to move applications from legacy to modern cloud-native technologies. We shall cover this later in this book. To conclude, bimodal IT, irrespective of falling out of favor, is still being practiced in some shape or form in most organizations. However, with digital transformation and cloud adoption gaining more and more momentum, most organizations are looking at standardizing Mode 2 by modernizing existing applications so that they're compatible with it. *Figure 3.5*, which is based on an IDC article called *The New Era of Digital Transformation*, categorizes digital business investments into three distinct buckets. Some of the investments under the foundational bucket are spent on building an efficient Mode 2 operating model, as well as modernizing some of the core business applications under Systems of Record and Systems of Differentiation:

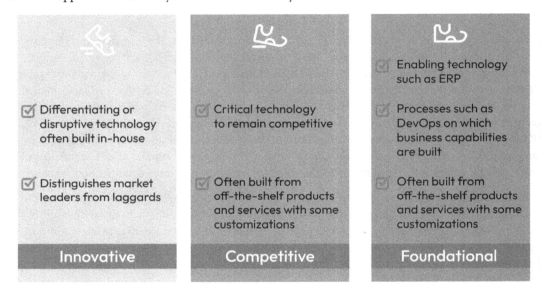

Figure 3.5 – Three categories of enterprise technology investments

We believe this journey will be in motion for the foreseeable future. Even with most apps and processes being moved to Mode 2, there may still be a small subset that will continue to operate in Mode 1.

Summary

In this chapter, we explained how Gartner's bimodal IT proposed a model to address the challenge of building and maintaining a diverse technology landscape with different modes of operations. The outcomes of this were improved agility, better risk management, greater innovation, increased collaboration, and an improved customer experience.

We also walked through the challenges and limitations of adopting bimodal IT in the ever-evolving enterprise IT landscape and why it is marked as inadequate to operate enterprise IT.

In the next chapter, we will discuss how hybrid cloud adoption is becoming a standard deployment model, as well as some of the key business and technical capabilities it provides.

Further reading (and listening)

For more information about the topics covered in this chapter, please refer to the following links and research on your own:

- Bimodal IT: How to Be Digitally Agile Without Making a Mess: `https://www.gartner.com/en/documents/2798217`

- 10 things you need to know about bimodal IT: `https://www.bloomberg.com/professional/blog/10-things-you-need-to-know-about-bimodal-it/`

- bimodal IT (bimodal information technology): `https://www.techtarget.com/searchcio/definition/bimodal-IT-bimodal-information-technology?_gl=1*1an3v4z*_ga*Mzg1NTgwNzY5LjE2NzY5NDk3MDA.*_ga_TQKE4GS5P9*MTY3Njk0OTY5OS4xLjEuMTY3Njk0OTc0MS4wLjAuMA..&_ga=2.15040502.1110145717.1676949700-385580769.1676949700`

- Who is a DevOps engineer?: `https://www.redhat.com/en/topics/devops/devops-engineer`

- Podcast (why we didn't agree on bi-modal IT): `https://www.buzzsprout.com/991558/436255`. Let's Bury Bimodal Thinking in Enterprise IT: `https://devops.com/lets-bury-bimodal-thinking-in-enterprise-it/`

- Why bimodal IT kills your culture and adds complexity: `https://www.cio.com/article/240840/why-bimodal-it-kills-your-culture-and-adds-complexity.html`

- Can the Elephant and the Cheetah be bedfellows?: `https://www.linkedin.com/pulse/can-elephant-cheetah-bedfellows-dhruba-ghosh/`

4
Approaching Your Distributed Future

Hybrid cloud and edge adoption has been growing steadily and is fastly becoming the standard deployment model for most organizations. In this chapter, we will explain what the hybrid cloud is, how it differs from multicloud, and the various internal and external factors that are shaping hybrid cloud adoption and its distributed future extending into edge computing. By the end of this chapter, we will have addressed the following:

- Key reasons why enterprises are moving toward a hybrid cloud and edge strategy

- External factors, from data sovereignty to ever-changing security and compliance requirements and AI/ML use cases that contribute to this journey

- How a more connected and distributed future with 5G/6G, IoT, and edge computing is playing a role

Let's start by revisiting what hybrid cloud is and how it differs from multicloud. A hybrid cloud is a diverse computing environment, where different types of infrastructures are used to run applications. The participating infrastructure leverages different environments, from public clouds, private clouds, on-premises data centers, and "edge" locations. Organizations choose to adopt the hybrid cloud to reduce costs, minimize risk, maximize flexibility, and support digital transformation efforts.

In some organizations, hybrid cloud and multicloud are sometimes used interchangeably. However, we believe they are two different things. This is important because it affects the dimensions of your operating model. The hybrid cloud includes different interconnected public and private clouds and the edge working together, sharing data and processes to run applications across the cloud. On the other hand, multicloud refers to an environment where an organization uses multiple public cloud services from different vendors to meet its needs. The main benefit of multicloud is the flexibility it delivers by letting organizations choose the best cloud services for their specific needs, rather than being locked into a single vendor's offerings through their proprietary APIs and the resulting highly vendor-specific engineering process of automation. For example, an organization can use data services from one cloud provider and AI solutions from another. In summary, leveraging multiple different types of infrastructure across public, private, and edge locations is called hybrid cloud, while leveraging public cloud services from different vendors is called multicloud. The following figure compares hybrid cloud and multicloud deployment models across two key perspectives:

HYBRID CLOUD	MULTICLOUD
Multiple deployment types – at least one public and one private – with some form of integration or orchestration between them	More than one cloud deployment of the same type (public or private), sourced from different vendors
Example: One private and one public cloud provider	Example: Two public or two private cloud providers

Figure 4.1 – Multicloud versus hybrid cloud

This chapter will cover the following main topics:

- Top reasons for hybrid cloud adoption
- The external factors
- Impact of 5G/6G, IoT, and edge computing

Top reasons for hybrid cloud adoption

As far as the reasons for hybrid cloud adoption go, we can classify them into two different categories:

- Business reasons
- Technology reasons

We shall discuss these classifications in detail here.

Business reasons

Hybrid cloud adoption doesn't always need to be a technical decision – in fact, most reasons for adoption or organically evolving to a hybrid cloud are not technical at all. This section focuses on the key business drivers in hybrid cloud adoption, which are as follows:

- Flexibility
- Cost optimization
- Reduce vendor lock-in risk
- Address everchanging security and compliance requirements
- Minimize business disruption

Let's get started.

Flexibility

A hybrid cloud allows organizations to choose the best deployment option and location for their specific needs – that is, a new cloud-native application can be deployed on one public cloud provider of choice and a data science workload can be deployed on another cloud that provides better data science capabilities. Similarly, certain workloads can be deployed in an on-premises data center for architecture, performance, data gravity, or security reasons. A hybrid cloud allows organizations to leverage a custom solution that meets their unique requirements, from the flexibility of an on-premise private data center to managed cloud environments and everything in between. Let's not forget edge deployments as well, which we will cover later in this chapter.

Cost optimization

A hybrid cloud helps organizations optimize costs by allowing them to leverage the existing technology investments made by the organization in their on-premise data center and related technologies. It also helps them optimize further as they can reduce cloud costs by leveraging the most cost-effective option for each workload, including network and storage costs, and better price negotiation leverage. *Please note that making deployment decisions with only cost in mind may not be the right strategy since other architectural elements need to be considered as well.*

Reduce vendor lock-in risk

This is the key business benefit hybrid cloud provides for most organizations. Leveraging a hybrid cloud capability that leverages a combination of different public cloud providers and on-premise and edge deployments reduces the dependency on any one particular vendor to a minimum. Organizations may still have single cloud provider dependencies for specialized workloads that are leveraged only from one specific cloud provider of choice. However, distributing the common applications around the hybrid cloud helps them operate independently and also be less impacted by business or technology changes happening with any single vendor. Another important aspect to highlight here is that every cloud provider is unique. From the types of infrastructure, application platforms, user interfaces, and APIs they provide, they differ from each other. Given this, organizations that build their entire application and infrastructure stack based on a single cloud provider end up with snowflake deployments that are not compatible with other cloud providers. This makes it extremely difficult, if not impossible, for organizations to migrate to a different cloud provider, move to a hybrid cloud model down the line, or repatriate workloads. Some of these challenges can be avoided by leveraging technologies and abstractions that are industry standards; we will take a closer look at this later in this book.

Address everchanging security and compliance requirements

Organizations are tasked with more and more regulatory and compliance requirements when it comes to data security and privacy. To make things complicated, most of these requirements are regional and fast evolving – for example, **General Data Protection Regulation** (**GDPR**) in the European

Union. This makes it difficult to partner with one vendor who can consistently meet all security and compliance requirements the organizations need to meet globally. Most organizations are now leveraging a combination of public cloud and on-premise data centers to address this challenge. We will talk about this in more detail later in this chapter.

Minimize business disruption

As organizations depend more and more on digital channels as the primary means to deliver their services, the availability of cloud environments that support these applications becomes critical. Cloud providers typically provide excellent global availability options with multiple data centers in the region and multiple regions across the globe to minimize cloud downtime. However, moving workloads from one region to another may violate security and compliance requirements that organizations need to fulfill and also impact application performance, which can decrease the overall customer experience. Adopting a hybrid cloud model with the ability to move applications/workloads across cloud providers in the same region helps organizations address any unforeseen cloud downtime issues without compromising regional security and compliance needs, as well as negatively impacting the customer experience.

In summary, flexibility, vendor lock-in risk, business continuity, customer experience, and regional security and compliance needs are some of the key business reasons for hybrid cloud adoption.

Technology reasons

Apart from these key business reasons, the following are some well-known technical reasons that favor a hybrid cloud deployment model:

- Legacy core business applications
- Edge computing

Let us discuss these technical reasons in the following sections.

Legacy core business applications

As we discussed in the previous chapters, many organizations still depend on legacy or traditional core business applications that are categorized under Systems of Record to deliver their core business services. These applications often have a lot of characteristics and limitations, which makes it extremely difficult, if not impossible, for them to be deployed in a cloud environment. Some of these characteristics are listed here:

- **Infrastructure dependencies**: Most core business applications were built before the cloud era and have a heavy dependency on the infrastructure they are deployed on, such as the operating system, processor architecture, and in a few cases specific storage and network types. These applications may also leverage direct hardcoded access to files also known as static file path and network resources, which makes them more bound to the infrastructure they are deployed on.

These infrastructure requirements and dependencies make these applications unsuitable for a cloud environment where the required snowflake infrastructure footprint and always-bound storage and networking layers are not available.

- **Security and compliance**: Core business applications were traditionally built for on-premise deployments and could only be accessed inside the corporate network. Given that they are protected from external threats and also secured by the physical and network security that is implemented, monitored, and improved upon by the organization to suit their regulatory and compliance requirements, you may think that physical and network security is also provided by the cloud providers. The move to cloud may not be possible, however, the key difference lies in the flexibility and choice in implementing your security and compliance standards. As an organization that's managing your data center, you have the freedom to implement physical and network security the way you want. For example, you can completely lock down the network from any inbound network connection or create fully air-gapped environments as part of your security posture. You can also implement strict data center access and reporting mechanisms that can be tailor-made to fit your security and compliance needs. This level of flexibility may not be provided by cloud providers at this point, though this may change as sovereign cloud notions   mature. With this freedom to implement physical and network security the way they want, core business applications are often built with fewer security features compared to modern cloud-native applications. They tend to communicate with backend systems on unencrypted open networks with minimal authentication and authorization checks. Business-sensitive data may also be stored in caches, logs, and other supporting systems. To add to the risk, most organizations also follow more lenient and mostly manual security and compliance review and approval processes for these systems that are deployed and accessed only inside the corporate network, given the minimal external security threats they have to address. Analyzing and addressing these issues to make the application ready for cloud deployment may simply be cost and time prohibitive for many enterprises, forcing them to leave these systems in their on-premise data centers and eventually work on retiring them with modern cloud-native applications down the line.

- **Extensions and integrations**: Core business applications, whether they are **commercial off-the-shelf (COTS)** applications bought from an ISV or core applications that have been built internally, are gradually   extended to add additional features as the business requirements evolve. These extensions typically involve integrating with other applications and data sources, creating a web of complex dependencies and communication patterns that work together to support a key business function. As time goes by, these applications and data dependencies will grow in complexity, eventually making all the participants tightly coupled with the rest of the applications and data sources.

- **Dated technology stack**: Depending on the age of the business application, the technology stack that's leveraged and the technical debt (the concept of accumulating technical deficiencies in software development projects as a result of prioritizing short-term solutions over long-term goals) that's incurred could simply make applications unsuitable for modernization or migration to cloud environments. We still see a few industries leveraging applications built on mainframe and client-server technologies to support their core business functions. These applications are there for a reason and will continue to provide business value in the foreseeable future. However, the longer they stay, the older they get and the more technical debt they incur, which makes them less suitable for hybrid cloud and edge deployment or modernizing them to a cloud-native architecture.

Edge computing

Edge computing is an important focus area for most enterprises thanks to the ever-growing IoT and edge use cases that are made possible with 5G/6G. From its early days in 2G, where few primitive IoT use cases were implemented, edge computing and IoT have grown leaps and bounds and are accelerating further as networks become more reliable and secured and devices become more capable. Gartner predicts that edge computing adoption will grow exponentially in the coming years. 75% of enterprise data in 2025 will be produced and processed outside the traditional on-premise data center or cloud. Edge computing adoption is prominent across all industries. Some of the top use cases and their corresponding industries are listed here:

- Smart cities in public services

- Telemedicine and patient health monitoring (eventually remote surgery) in healthcare

- Real-time inventory management and theft monitoring in retail

- Safety systems, assembly line monitoring, and analysis in manufacturing/logistics

- Self-driving vehicles in transportation

In summary, edge computing and the ability to produce, process, and manage huge volumes of data outside the traditional data center and cloud environments have triggered a new wave of applications and deployment models that do not fit well with traditional on-premises data center and cloud deployment models. Even though we like to consider edge as a single deployment unit, in reality, there are at least three different deployment models or infrastructure layers that exist in edge deployments:

- **Device edge**: Device edge refers to the device itself – for example, sensors, cameras, microcontrollers, and so on. These devices come with a minimal computing infrastructure that can be utilized to deploy meaningful applications to monitor, process, and store or send information to the backend systems. The device edge is often used for real-time data processing and device control with minimal latency and bandwidth requirements.

- **Far edge**: Far edge refers to a model where computing infrastructure is deployed near the end user or devices and further away from the core data center or cloud. Far edge is typically deployed within the same building or campus. The typical infrastructure components in far edge devices can include routers, switches, and other networking equipment, as well as servers or other computing devices that are located in a local data center. The most common use case for far edge deployment is to build atomic processing units that can operate independently, even when connectivity to the core on-premise data center or cloud is unavailable – for example, a back office server in retail store point of sale systems, ticketing and reservations systems in train stations, and so on. Increasingly, far edge is also often used for more advanced data processing, analytics, and storage to address reduced latency and bandwidth requirements by processing and storing data closer to the source. One trend to observe is that as end devices become more and more capable with better processing power and improved connectivity, the more powerful and autonomous the far edge deployment use case will be, helping organizations implement more advanced edge solutions with reduced infrastructure cost and improved security.

- **Near edge**: Near edge refers to a data center that is located farther away from the end user or device and nearer the core data center or cloud. Near edge is typically deployed in a regional data center or cloud environment. Near edge can support multiple far edge deployments, which, in turn, support multiple device edges. Near edge deployments are used for more intensive data processing, analytics, and storage applications, such as AI and machine learning applications.

The following figure shows how device edge, near edge, and far edge are deployed in an enterprise context:

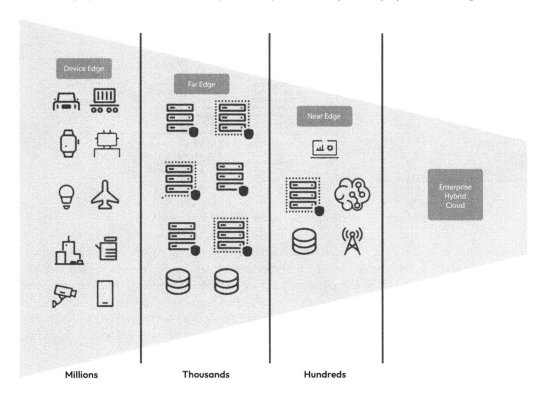

Figure 4.2 – Edge deployment models

A quick note to point out here is that the telco industry typically sees this representation differently. Edge deployments can be represented as follows:

- **Device edge**: Standalone (non-clustered) systems that directly connect sensors via non-internet protocols – for example, a coffee shop or your local restaurant

- **End user premises edge**: This includes retail stores, trains, or even houses or cars

- **Service provider edge**: Edge tiers located between the core or regional data centers and the last mile access are commonly owned and operated by a telco or internet service provider

- **Provider or enterprise core**: Traditional "non-edge" tiers owned and operated by public cloud providers, telco service providers, or large enterprises

The following figure shows how the edge computing landscape is represented by remove Red Hat:

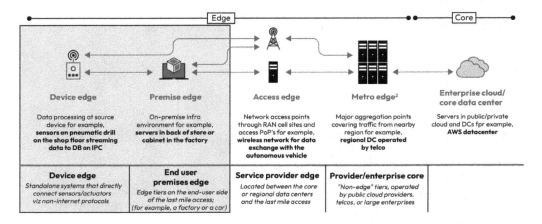

Figure 4.3 – Red Hat edge architecture representation

Irrespective of the representation, the ability of modern smart devices to participate in data creation, processing, and analysis has extended the reach of applications, from core data centers and cloud to edge devices to near edge and far edge deployments. Edge deployments, especially the device edge and far edge, will require different if not specialized hardware that lends itself automatically to a hybrid cloud deployment model. Edge may also add additional complexities to the ongoing maintenance and life cycle management of the infrastructure. We will spend more time on this later in this book, especially in *Chapters 5* and *6*.

The external factors

External factors that are required or mandated play a significant role in enabling organizations to implement a hybrid cloud architecture. Some of these external factors are as follows:

- Regional compliance requirements
- Infrastructure limitations
- Mergers and acquisitions

Regional compliance requirements

Regional compliance (and security) requirements are the most common and well-known reason for organizations to embrace a hybrid cloud deployment model. One such requirement is the data sovereignty requirements that differ from region to region. Here are some examples of data sovereignty requirements:

- **European Union**: The data sovereignty requirement for the EU is called the GDPR. The GDPR requires that the personal data of EU citizens is stored and processed within the EU or in a country that the EU has deemed to have adequate data protection laws.

- **Russia**: The Personal Data Law of Russia requires that the personal data of Russian citizens must be stored and processed within Russia.

- **United States**: The data sovereignty requirements differ from state to state in the United States. The most comprehensive is the **California Privacy Rights Act** (**CPRA**), formerly known as the **California Consumer Privacy Act** (**CCPA**), which mandates that organizations that collect and process personal data of California residents must provide certain privacy rights to those residents. CPRA requires for-profit organizations above a certain size to collect customers' consent before selling their data. It also requires organizations to provide an option for the customers to prevent their data from being sold with a *Do not sell my personal information* option.

- **India**: The Personal Data Protection Bill of India requires that personal data must be processed within India or in a country that the Indian government deems to have adequate data protection laws. This law is still in its infancy and additional requirements are expected to be added in the coming years.

- **China**: Similar to the countries mentioned previously, the Cybersecurity Law of China requires that critical data infrastructure operators must store personal and important data in China and conduct security assessments before transferring data abroad.

- **Singapore**: Singapore has a **Personal Data Protection Act** (**PDPA**) in which organizations are subject to the Transfer Limitation Obligation. An organization must not transfer personal data to a country or territory outside Singapore except as per the requirements prescribed under the PDPA to ensure that the transferred personal data will be accorded a standard of protection that is comparable to that under the PDPA.

There are also strict data sovereignty requirements for Australia and New Zealand.

A global organization that operates across multiple regions may have to address many such regional data sovereignty requirements that will encourage them to adopt a hybrid cloud deployment model with many local regional deployments to fulfill the local data sovereignty laws. At this point, it is difficult to find a single cloud provider who can address all the local requirements globally. However, organizations are advised to utilize sovereign cloud to address key regional data sovereignty requirements. This may be a viable option in the far future, especially for organizations that operate only in key global regions where sovereign cloud options are present.

Apart from the regional compliance requirements, other cultural aspects may come into play as well. For example, I used to consult for a healthcare company in North America that had implemented a risk management program to distribute drugs to ensure that the patients are not pregnant during the treatment cycle to address birth defect risks associated with their drug. They did this by making the patients and doctors go through a risk management process every month before each prescription. This process involved asking personal questions about a patient's use of birth control and pregnancy test results. This program was first launched in North America, where patients of all ages, starting with age 15, were asked to respond to this survey either online or with the aid of a nurse before the prescription was written. However, when the company got approval to launch the product in Asia, the same program did not seem to work. The main reason for this is that in most Asian countries, older parents and sick family members are tended to by family members who often help the patients with these monthly surveys. Some of the questions related to birth control usage and pregnancy test results are too sensitive to ask the patient, especially parents, and had a negative influence on the usage of this life-saving drug. This forced the organization to come up with different programs for different regions based on subtle cultural aspects. To address this, rather than building a single application with multiple different variations, the organization decided to deploy regional applications in the respective regions, some on the public cloud and some on on-premise data centers, depending on cost, cloud availability, and regional compliance requirements.

In summary, external factors play a significant role in defining the cloud architecture of most organizations. It could be technical, compliance, cultural, or a combination of more than one factor. Having the ability to respond to these requirements or challenges and being flexible to change the technology and business response is critical to the long-term success of the organizations, especially when they expand globally.

Infrastructure limitations

Infrastructure limitations may have an impact on an organization's deployment model as well. Depending on the infrastructure requirements, organizations need to deliver meaningful services to their customers and the regions where they provide services. Organizations may have to go with a distributed regional deployment model either on the public cloud or on-premise (private cloud) or a hybrid cloud model. Given the maturity of the cloud ecosystem in certain regions, most organizations are likely to go with a hybrid cloud model. One such example is a global bank I supported as an architect in the recent past. Even though they are based out of Western Europe, they have a significant presence in Southeast Asia and Sub-Saharan countries in Africa. To address the network limitations in Sub-Saharan and some Southeast Asian countries, they decided to go with a hybrid deployment model with a combination of regional cloud deployments on a prominent cloud provider and in-country private cloud deployments for Sub-Saharan and some Southeast Asian countries. For example, there was a regional deployment in the UK with a prominent cloud provider for Western Europe, a regional cloud deployment in Hong Kong for a few Southeast Asian countries, and in-country private cloud deployments for the remaining Southeast Asian countries and for Sub-Saharan countries. This model helped them achieve their business goals with improved customer satisfaction, irrespective of higher ongoing operational costs in maintaining multiple data center footprints in their hybrid cloud environment.

Mergers and acquisitions

Mergers and acquisitions (**M&A**) is a common approach many organizations take as part of their growth strategy. Here are some of the main reasons for M&A:

- **Increase market share**: Mergers and acquisitions are used by organizations to increase market share by acquiring competitors or complementary businesses. It can also help organizations improve their position in the market.

- **New market access**: M&A can provide access to new geographic or product markets that may take organizations a long time to establish and grow organically, thus helping them reduce risk and improve profitability.

- **Portfolio diversification**: Organizations use M&A to diversify their business. This is done by acquiring other organizations with complementing business models in the same or similar industry. Organizations can also leverage the newly acquired business to build unique value-added features/capabilities to differentiate from their competitors further.

- **Threat mitigation**: M&A is also used to address new threats emerging in the industry that may potentially disrupt the organization in the near future. By acquiring this budding competition early in the cycle, organizations can address the emerging threat efficiently and also use the acquired business to fuel revenue growth in the mid to long term.

M&A often brings in a different IT landscape that may not be compatible with acquiring an organization's existing IT landscape. This increases the probability of organizations adopting a hybrid cloud model post-M&A. Depending on the organization's business model and technology diversity and complexity, a hybrid cloud model may be the only viable option and will continue to stay that way for the foreseeable future, with both organizations operating independently with minimal overlap.

Impact of 5G/6G, IoT, and edge computing

You may be wondering why the impact of 5G/6G, IoT, and edge computing has been put together in this section. We believe they are logically connected in aiding the adoption of hybrid cloud and in delivering edge use cases across the industries. We need to classify edge computing into two different eras:

- Traditional edge computing (pre-5G)
- Modern edge computing (post-5G)

The important differences between these eras are the explosion of new use cases and architecture patterns that modern edge computing can support thanks to the reliability, data transfer rate, and security that a 5G/6G network can provide, along with the increased processing capabilities of smart devices. In this section, we will focus on modern edge computing (post-5G).

For modern edge deployments to be successful, we need the following:

- A reliable high-speed and secured network that's provided by the 5G/6G network

- Smart devices capable of data collection and processing (and storing), which is the core of IoT

- Ability to process, report, and analyze data close to the source via edge computing

From a hybrid cloud deployment perspective, 5G/6G, IoT, and edge computing all play a role:

- **5G/6G**: 5G/6G technology provides faster and more reliable connectivity, which helps organizations improve the performance and reliability of their edge environments. With 5G/6G, organizations can move large amounts of data between on-premise infrastructure, cloud-based infrastructure, and edge infrastructure, including edge devices, more quickly and with lower latency, reliably and securely. This can help enable real-time data analysis and decision-making in a hybrid cloud environment and also provide the capability to build broader and farther-edge deployment models that were not feasible with prior-generation networks.

- **IoT**: IoT involves connecting devices to collect, share, and analyze data. This data can be processed in a hybrid cloud environment, with some data processed locally on edge devices and some data processed in the core data center or cloud, and some in-between at near edge or far edge locations. IoT devices can also help improve the performance and reduce the cost of hybrid cloud environments by reducing the amount of data that needs to be transferred to the cloud with the capability to process them locally and also make smart decisions in the case of adversity. This ability to capture, share, analyze, and process data and make smart choices in the case of adversity has created a lot of interesting use cases, from remote surgery to smart surveillance. Needless to say, 5G/6G networking and edge computing play a critical role in implementing these futuristic and impactful use cases, which are gaining more and more popularity.

- **Edge computing**: Edge computing involves processing data locally on edge devices or at near edge /far edge infrastructure rather than just in the core data center or the cloud. This can help reduce latency and improve the performance of hybrid cloud environments. By processing some data locally on edge devices and some data in the cloud, organizations can create a more efficient and effective hybrid cloud environment.

Advancements in these three areas have created more opportunities and business needs for organizations to implement use cases that do not fit well with single-cloud or multicloud deployment models. As we saw earlier in this chapter, across industries, more and more edge use cases are being implemented thanks to the accelerated innovation in these areas.

Summary

In this chapter, we explained how the innovation in the fields of 5G/6G, IoT, and edge computing is creating new business value across industries, which is creating a need for a hybrid cloud and edge deployment model. We also discussed some of the top use cases it supports and how some regional data sovereignty and compliance requirements can influence the adoption of hybrid cloud and edge computing. We also discussed the external factors that can play a role in hybrid cloud adoption, including M&A.

In the next chapter, we will discuss how organizations can build a hybrid cloud and edge operating model that suits their business needs.

Further reading

For more information about the topics that were covered in this chapter, please refer to the following links:

- Hybrid cloud strategy and the importance of on-premises infrastructure : `https://www.ibm.com/blogs/systems/forrester-study-hybrid-cloud-strategy-and-the-importance-of-on-premises-infrastructure/`

- What Edge Computing Means for Infrastructure and Operations Leaders: `https://www.gartner.com/smarterwithgartner/what-edge-computing-means-for-infrastructure-and-operations-leaders`

- The Cost of Cloud, a Trillion Dollar Paradox: `https://a16z.com/2021/05/27/cost-of-cloud-paradox-market-cap-cloud-lifecycle-scale-growth-repatriation-optimization/`

Part 2: Building a Successful Technology Operating Model for Your Organization

In this part, you will learn in detail about the building blocks for a distributed operating model across the cloud and edge, followed by a real-life use case, and how this organization built its Distributed Technology operating model in a hybrid multi-cloud and edge context. It then connects a real-world architecture, component design, and implementation with the previously developed operating model. Finally, the part concludes with a summary chapter that highlights all the key knowledge from the book and also introduces a few additional takeaways for you.

This part has the following chapters:

- *Chapter 5, Building Your Operating Model for the Distributed Future*
- *Chapter 6, Your Distributed Technology Operating Model in Action*
- *Chapter 7, Implementing Distributed Cloud and Edge Platforms with Enterprise Open Source Technologies*
- *Chapter 8, Into the Beyond*

5

Building Your Distributed Technology Operating Model

In this chapter, we will be putting pen to paper by walking you through the process of creating an operating model. By the end of this chapter, and as part of building your operating model, you will have covered the following aspects:

- How to assemble your stakeholders and how to group them in relation to your operating model tasks and the tools you can use for stakeholder management
- Practices that you can employ to run workshops and achieve the desired outcomes
- A way to slice and dice a multidimensional task, such as creating a distributed technology operating model

The following topics will be covered in this chapter:

- Step-by-step instructions to build your operating model for the distributed future
- Using open practices to build an open culture and create ownership
- Stakeholder management

Building your operating model for the distributed future

It is common to see organizations implementing different operating models for different environments. That again is mostly due to different change cadences, data sensitivity, and access requirements.

The following different environments can be observed:

- Workplace/modern workplace – for example, Office suites, document storage, knowledge management, and policies
- Backoffice environments such as service management
- Revenue-generating, customer-facing, front-office environments

While the approach provided in this book can be used across all environments, the dimensions examples are geared toward cloud and edge native **customer experience (CX)**, improving revenue-generating services.

In the previous chapters, we have used the term **hybrid cloud and edge** or **hybrid multicloud and edge** for clarification purposes. From now on, we will refer to hybrid, multicloud, and edge concerns generally as distributed cloud or cloud platforms within the context of creating your technology operating model.

Enterprise agility and operational effectiveness depend largely on the responsiveness, scalability, and resiliency of the digital infrastructure used to enable mission-critical applications, data operations, and connectivity for customers, partners, and employees. Resiliency and agility are particularly crucial to business success in periods of disruption and uncertainty. Examples include disrupted supply chains, rising inflation, geopolitical tension, price spikes, a pandemic, or climate change. And this is why we advocate for an operating-model first approach, not a technology-led or even a use case-led approach to embark on a cloud or edge journey. Influencing factors can come from many different directions, with many aspects to be aware of. We group these aspects into dimensions. The scope of those dimensions makes this task manageable.

It's time to start creating our best-fit distributed cloud and edge operating model. As we've learned, there is no right or wrong, but rather a good starting point to iterate on as your organization learns and optimizes through different operating model iterations. We also have McKinsey as a big fan of an iterative approach to operating model development, and Forrester is in agreement that you have got to build your own and not work off prefabricated templates. If you get stuck, you can always have a peek at those prefabricated templates to see what dimensions and angles they take. However, across different templates, you will encounter a lack of a consistent structure or approach. Plus, you will have to adopt those templates for your organization's needs. So, with all this taken into account, we are well-placed to start building a fit-for-purpose operating model. And in this instance, the practitioners and management consultants agree.

So, let's press ahead.

Starting at the end

Before we select our dimensions and order them with regard to our organizational priorities, let's have a look at what we want the outcome to be and the process we will follow to arrive at the outcome.

The following road map shows the main steps we will follow to create our operating model dimensions for each operating model stream and, subsequently, our complete technology operating model for our distributed future:

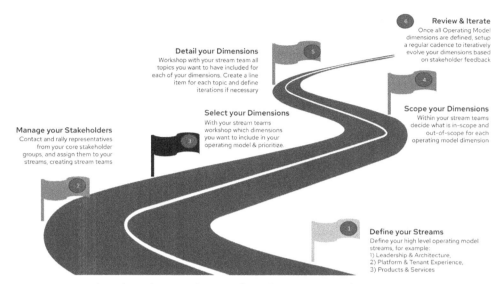

Figure 5.1 – Flow chart showing the steps from the start to your finished operating model

To better select the right stakeholder groups for the respective topics, we propose that we subdivide the operating model concerns into different streams. The reason for this is that not all topics are relevant for all teams or people. For example, while topics such as DevOps and the development team's structure are relevant and interesting for developers, cost-center setup, project accounting, or how to move organizationally from project to product budgeting might not be, or at least to a lesser extent. Plus, we all have a day job and want to feel productive. So, while we should communicate and share the outcomes of workshops across teams, not everyone needs to be involved in every workshop in person. It's also important to document decisions, just like we used to do in the good ol' architecture decision log. This allows us to revisit the *why* behind decisions if needed.

We are proposing that we divide and conquer by utilizing the following streams:

- Leadership and (Enterprise) Architecture
- Platform and Tenant Experience
- Products and Services or Applications and Workloads, depending on what you are building

A completed operating model dashboard can be seen in *Figure 5.2*. On the left-hand side, we have the streams mentioned previously and the dimensions that the streams own. Because all the operating model definition work here has been completed (which is probably never the case in reality), no dimension is currently **BLOCKED**, in the **BACKLOG**, a **work in progress** (**WIP**), or in a **DEFINED** state. All operating model dimensions are currently **IN USE**, meaning they are currently actively applied or used. As soon as this happens, feedback loops are activated and new change requests to further optimize your operating model can come in.

Successful organizations keep refining their operating model continuously to keep it relevant, effective, and ready to support the changing needs of the business without building up change inhibitors due to technical or process debt. New requirements can arise from new technologies becoming available, new practices promising better outcomes, internal process improvements, new security or compliance controls being mandated, parts of processes being optimized, or better pricing available elsewhere. This operating model's **sensing capability** is what triggers the ongoing and interactive improvement of our operating model and what keeps it alive:

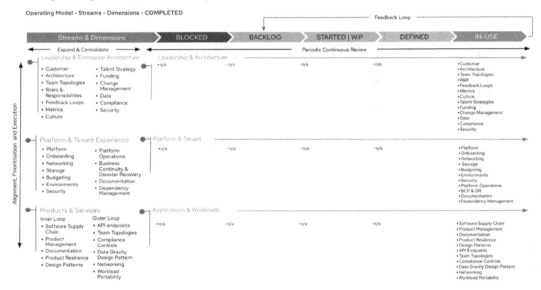

Figure 5.2 – Operating model dashboard showing an in-time snapshot of a completed operating model, just before new feedback loop-based requirements hit

The one-page dashboard depicted here is a good way to communicate what's going on across the different streams concerning the operating model dimensions and helps align and synchronize across streams and dimensions. It can also show backlog items, the operating model **dimension's design ownership**, and the immediate next steps. That said, it's not mandatory to stick exactly to this style. If you have dashboard styles that your organization is already familiar with, you can use them to communicate as well.

The dimensions themselves provide target and transition states, guiding principles and guardrails for managers and individual contributors alike. As a quick reminder, guiding principles are soft boundaries, while guardrails are hard boundaries. For example, guiding principles provide broad, high-level guidance and serve as a foundation for decision-making, while guardrails are more specific and operational, providing boundaries and constraints to ensure compliance and mitigate risks. Now, while guiding principles sound simple, it can be difficult to run a workshop to agree on *guiding principles*. You can find yourself sandwiched between "motherhood" statements (we leverage the goodness of the cloud), empty, meaningless, or confusing phrases (cloud-first, anyone?), aspirations, and actual guiding principles. A good guiding principle describes the intention of the principle and the implication to provide a clear understanding of why and what the organization is trying to achieve through the principle. This then becomes a clarity-providing guiding principle as it provides stakeholders with the spirit of the principle as well.

Here's an example: "*We use proprietary cloud services over building our infrastructure services. This means that we standardize the services within a single cloud environment and consume its services. The consequences we accept are that we will have little reuse and the same amount of engineering effort, skills development, and compliance work to do when the move to a multicloud environment becomes necessary. The reason we are doing this is because our primary aim is to prove that there is a market for our offering as a start-up.*"

I like manifestos such as the Agile manifesto because they compare different aspects and by doing so provide context and set a priority that helps clarify what that principle means and makes it more actionable. A good example is "individuals and interactions over processes and tools." You can use priority sliders (see OPL) to establish foundational guiding principles to determine trade-offs across cost, security, automation as code, and so forth.

The core of communicating and sharing your operating model is documentation and code that can be accessed by anyone in a known place. For documentation, I prefer a wiki or a Git repository rather than documents as it's easier to always access the latest and greatest version. We certainly do not want outdated documentation floating around. Other parts of your live operating model infrastructure include code repositories, container registries, and self-service portals.

Your operating model assets shape and guide the actions, decisions, and behaviors of an individual or team concerning the application and execution of your distributed operating model. They serve as a framework for decision-making and help ensure that actions are aligned with the desired outcomes. Ultimately, they empower your people and teams. Here are some examples of manifesto-style principles and guardrails in a distributed cloud context:

- Workload portability over cloud provider proprietary services, team incentives over individual incentives or remote data access over ETL, physically copying data, build over buy, reuse over recreate, open source over proprietary solutions, and product over project budgeting (guiding principles)
- Guardrail examples ensure that the build pipeline does automatic software bill-of-material vulnerability checks before deploying into production, that all data access is only available through the enterprise data mesh, or that all infrastructure requests can only be made through the internal developer portal

The right principles and guardrails can be used to create a positive and productive work environment, establish a clear sense of direction, promote ethical behavior, empower teams, and support fast decision-making.

Other more playful direction-setting approaches I've seen are that if your company's name starts with D, it could be *We Want to Be the D in "GRAND"*, where the other letters refer to industry players you want to be similar to, such as Google, Red Hat, Amazon, Netflix, and your organization.

When the excitement around creating your operating model for your distributed future has slowed down you need to have an owner or home in place. Ideally, this is a product owner-like persona with a budget and **cross-line of business** (**cross-LOB**) remit and sponsorship. Like with many things in business, if you get it right, then this will be an easy endeavor. If your operating model fails to deliver value to the business, then it will be hard. Keep this in mind when you work on the customer, funding and budgeting, onboarding, and success criteria topics. If product management is not something your organization is familiar with, then enterprise architecture (not IT enterprise architecture) is a good home too. A function with an enterprise-wide scope usually offers a good place of ownership. If that's not available, then other owners can be your digital teams or your CIO or CTO office, or your chief operations officer's office. We recommend to stay away from teams that have a narrowly scoped product innovation agenda only. An operating model is less aligned with narrow-scope thinking and the related MVP-focused, fail-fast and move-on mindset.

Managing your stakeholders

People are everything. To get the right alignment and the best-fit operating model for your organization, you need to ensure the right stakeholders are assigned to the streams and are actively participating so that the voices of people that have insight and vision provide input. We won't look at stakeholder management in depth. However, we do recommend creating a stakeholder list as a starting point. This way, you will get an idea of who you need to work with to get things done, and you also won't forget to invite key people to your workshops.

Existing organizations often already have org charts that you can use as a starting point. Typically, the **project management office** (**PMO**) or enterprise architecture function are good places to ask if you can't find it. Furthermore, you can look at your application portfolio matrix to find the workload owners, business and technical support staff, as well as who is looking after the underlying infrastructure, and finally the users as the "customers" of those applications. Then, you have the project and program portfolio with planned and in-progress activities, which also provides you with a good view of what's coming up and the future stakeholders of your operating model. And you can always turn to the Open Practice Library and run an impact mapping exercise that maps out all stakeholders that are associated with the stated goal, which could be something along the lines of "creating the best-fit distributed cloud and edge operating model."

Once you have your stakeholder list, there are several different categorizations for stakeholder matrices you can find. They are usually 2x2 matrices, where the axes have a low/high unit of measurement. They are as follows:

- Power and Interest
- Commitment and Importance

- Power and Support
- Support and Importance
- Impact and Influence

There are probably many more. The most useful ones for developing our operating model are Power/Influence and Impact.

While Influence is about how powerful a stakeholder is to steer your organization toward a new direction, set priorities or provide or limit funding, Impact is about how much a stakeholder is impacted by your cloud operating model. And that's our cloud operating model customer. And we dig customers. A lot.

Once you have matrixed your stakeholders, you need to use this information to decide who you need to invite to participate in your workshops. Individuals and teams with high influence and low impact can be represented in kick-off workshops if the number of participants becomes too high. Within those workshops, you will also find out who should be informed or consulted about other streams or decisions for other dimensions. To keep track of this, you can create a **Responsible, Accountable, Supporting, Consulted and Informed** (**RASCI**). A RASCI model specifies who from your stakeholders is Responsible, Accountable, Supporting, Consulted or merely informed for specific work streams and the associated tasks.

"When does the group size become too large?" you might be wondering. This is a good but hard-to-answer question. We covered Dunbar's number previously, but the book *Team Topologies* by *Manuel Pais* and *Matthew Skelton* dives into this further if you are interested. Here's a rough guideline that you can refer to:

- **Five people**: The number of people you can hold a close relationship and working memory with
- **15 people**: The number of people with whom you can experience deep trust
- **50 people**: The number of people we can have mutual trust with
- **150 people**: The number of people whose capabilities we can remember

Now, while it's nice to know there are corporate realities, such as stakeholder voices that need to be heard and people who need to be involved for change management reasons, as well as SMEs that need to be involved, if your stream team is getting too large to be workable, then you can create sub-stream teams that own specific dimensions. I just want you to consider the admin overhead and synchronization efforts before you go down that path. The recommendation is not to create any further team divisions, and to avoid anything beyond 150 people working on a single operating model stream.

High-impact and high-influence stakeholders in your stream should be prioritized as those that attend all your workshops; so should low-influence but high-impact ones. If the number of attendees becomes too much, my recommendation is to focus on the high-impact stakeholders and teams to ensure uptake, ownership, and the best possible CX while keeping noise and detractions low.

The following is an example of what a RASCI could look like.

Sometimes, you also see a RASCI, which includes additional stakeholders who *support* the implementation, meaning those who are not directly responsible but help out once in a while. You should use this to proactively communicate resource requirements.

RASCI can be furthermore explained as follows:

- **R – responsible**: The person responsible for carrying out the entrusted task, who is doing the work

- **A – accountable**: The person that is accountable for the outcome and what has been done

- **S – support**: The person that provides support during the implementation of the activity/process/service

- **C – consulted**: The person that can provide valuable advice or consultation for the task

- **I – informed**: The person that should be informed about the task's progress or the decisions in the task

Once you have communicated and understood those definitions, you can create your operating model-related RASCI chart. Here is an example. I like to call team members out by name as it creates both accountability and a sense of ownership:

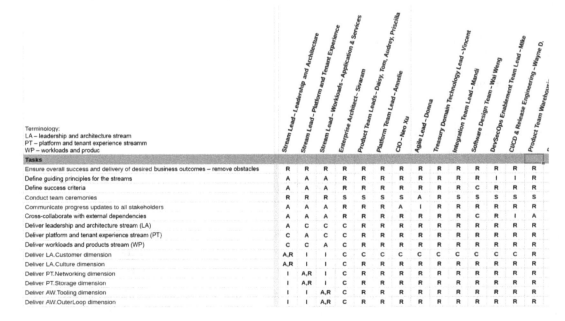

Terminology: LA – leadership and architecture stream PT – platform and tenant experience streamm WP – workloads and produc	Stream Lead – Leadership and Architecture	Stream Lead – Platform and Tenant Experience	Stream Lead – Workloads – Application & Services	Enterprise Architect – Sivaram	Product Team Leads – Daisy, Tom, Audrey, Priscilla	Platform Team Lead – Amelie	CIO – Neo Xu	Agile Lead – Donna	Treasury Domain Technology Lead – Vincent	Integration Team Lead – Mandi	Software Design Team – Wai Weng	DevSecOps Enablement Team Lead – Mike	CI/CD & Release Engineering – Wayne D.	Product Team Warehous...
Tasks														
Ensure overall success and delivery of desired business outcomes – remove obstacles	R	R	R	R	R	R	R	R	R	R	R	R	R	R
Define guiding principles for the streams	A	A	A	R	R	R	R	R	R	R	R	I	I	R
Define success criteria	A	A	A	R	R	R	R	R	R	R	R	C	R	R
Conduct team ceremonies	R	R	R	S	S	S	S	A	R	S	S	S	S	S
Communicate progress updates to all stakeholders	A	A	A	R	R	R	A	I	R	R	R	R	R	R
Cross-collaborate with external dependencies	A	A	A	R	R	R	R	R	R	R	C	R	I	A
Deliver leadership and architecture stream (LA)	A	C	C	C	R	R	R	R	R	R	R	R	R	R
Deliver platform and tenant experience stream (PT)	C	A	C	C	R	R	R	R	R	R	R	R	R	R
Deliver workloads and products stream (WP)	C	C	A	C	R	R	R	R	R	R	R	R	R	R
Deliver LA.Customer dimension	A,R	I	I	C	C	C	C	C	C	C	C	C	C	R
Deliver LA.Culture dimension	A,R	I	I	C	R	R	R	R	R	R	R	R	R	R
Deliver PT.Networking dimension	I	A,R	I	C	R	R	R	R	R	R	R	R	R	R
Deliver PT.Storage dimension	I	A,R	I	C	R	R	R	R	R	R	R	R	R	R
Deliver AW.Tooling dimension	I	I	A,R	C	R	R	R	R	R	R	R	R	R	R
Deliver AW.OuterLoop dimension	I	I	A,R	C	R	R	R	R	R	R	R	R	R	R

Figure 5.3 – A RASCI matrix example

Your resulting RA(S)CI and stakeholder matrix also informs your communications plan going forward, telling you who you need to keep across updates, progress, and invites to showcase events.

While you are implementing the operating model so that it's in its defined target state (To-Be), you should keep your stakeholders abreast of the changes and how they impact them:

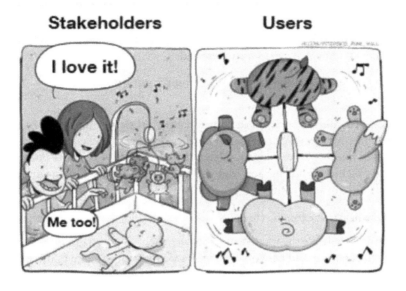

Figure 5.4 – A reminder that it's easy to miss a perspective affecting User eXperience (UX)

Once you have assigned your stakeholders to specific streams, it is time to start organizing your kick-off workshop. If time permits, we recommend having one-on-one conversations or documented interviews upfront with the stakeholders to get their undistracted and hopefully honest views. Group settings can sometimes make people shy or political and not offer what they are really thinking. The Open Practice Library's practices help create an open and collaborative environment but nevertheless, we still recommend one-on-one meetings and interviews as the first step so that key stakeholders feel included, as well as so that you can understand their vision and thought process, which can influence the resulting operating model.

Selecting your dimensions

The goals of all your operating model workshops are as follows:

- To create a collaborative environment and psychological safety to enable open sharing
- To select and prioritize the dimensions for your streams
- To define the success criteria for your streams and dimensions

- To define all relevant aspects of your dimensions (scope) and then associate your guiding principles and guardrails with them

When you organize your workshops, think about whether you want the group to converge or diverge. Based on your intended outcome, you let the participants discuss first and brainstorm second (converge – in case you want a set of coherent output) or silently brainstorm first and discuss second (diverge – if you want many different ideas). Talking with others or reading preparatory material primes the participants' brains. But never forget that the magic happens when people share and talk to each other since common understanding and synergies are created.

The workshop exercises described in the following paragraphs can be spread over several different workshops.

Creating an environment for collaboration and open sharing is the most important thing to do. Good facilitators can read the room very well (even in a virtual environment) and can schedule ad hoc breaks or quick games when needed, know how to quieten the **highest-paid person's opinion (HIPPO)**, and keep people feeling heard and engaged. You will fare better with a self-managing group; hence, we propose two exercises to start with:

- **Ice breaker**: To help participants get to know each other and discover similarities among themselves.
- **Social contract**: To define the wanted and unwanted behaviors as a group. Each team member has to sign that contract and hence can be held accountable to adhere to the social contract.

Before you get started with your dimensions, we recommend another open practice called **Definition of Done**. This means that the criteria that need to be met to consider a dimension as complete, done, or *defined* are agreed upon across the team. These collaboratively created criteria can then be maintained, enforced, and referenced by the team ongoingly.

You are now ready to work on the stream dimensions and success criteria that are the most suitable for your organization. We have provided our recommendations here for you to consider. Based on whether you want the group to converge or diverge determines if you want to share the recommended dimensions upfront or establish your own and just reference check with what we suggest after to see if you missed anything.

To find your success criteria and operating model dimensions, you can use the following practices:

1. Start by brainstorming, such as brainwriting or silent brainstorming, 10 for 10, and 1-2-4-all:

   - Silent brainstorming and brainwriting or 10-for-10 are quick ways to generate a lot of diverse ideas while minimizing bias introduced by others verbalizing their suggestions (diverge)

   - 1-2-4-all are brainstorming exercises that are based on collaboration and hence converge ideas

2. Group similar items together and find a name for the dimension that best incorporates all the different angles.

3. Prioritize the dimensions. This helps you work out dimensions based on the highest need. Prioritization is also helpful for clarifying the purpose of the dimensions. You can run a simple voting exercise, such as using dots on sticky notes, or priority sliders, as per the Open Practice Library.

The workshop's length depends on many things, such as stakeholder availability, the scope of topics, workshop frequency, how well the team is working together, and how familiar the participants are with the practices and the desired outcome. The trick is to find a good productive balance. Don't make people sweat for 8 hours straight, but also set enough time aside to ensure the necessary conversations and discussions can happen.

Once your dimensions have been established but not yet prioritized across your streams, your dashboard should look similar to this:

Figure 5.5 – Operating model dashboard with selected dimensions

On the left-hand side of the dashboard, you can see the operating model streams and the operating model dimensions that were selected by the stream teams:

Streams & Dimensions

← Expand & Consolidate →

Leadership & Enterprise Architecture

- Customer
- Architecture
- Team Topologies
- Roles & Responsibilities
- Feedback Loops
- Metrics
- Culture

- Talent Strategy
- Funding
- Change Management
- Data
- Compliance
- Security

Platform & Tenant Experience

- Platform
- Onboarding
- Networking
- Storage
- Budgeting
- Environments
- Security

- Platform Operations
- Business Continuity & Disaster Recovery
- Documentation
- Dependency Management

Products & Services

Inner Loop
- Software Supply Chain
- Product Management
- Documentation
- Product Resilience
- Design Patterns

Outer Loop
- API endpoints
- Team Topologies
- Compliance Controls
- Data Gravity Design Pattern
- Networking
- Workload Portability

Figure 5.6 – Zoomed-in view of the operating model dashboard

Subsequently, you must utilize your open collaborative environment and your social contract to get the group to agree on the scope, target and transition states, the guiding principles (soft boundaries) and the guardrails (hard boundaries) for each operating model dimension.

Scope simply means what aspects of a dimension are in scope and out of scope. For example, in the networking dimension in the workload and application stream service mesh, layer 7 application interconnect, SR-IOV might be in scope, but global traffic management might be out of scope.

Now that we have established working relationships within the stream team, we progress through the forming, storming, norming and performing team-building phases. Here, we must vote, prioritize, and, most importantly, have conversations around different topics, angles, view points, and so on. This leads to a shared language and shared understanding among the participants.

Scoping your dimensions

The next work item is to determine the scope. The scope is not only important to have a view on who and what subject matter expertise you need, but it also quickly shows you what topics have dependencies in other streams and require input and synchronization.

After the kick-off workshop, I recommend that you run the next workshop within your stream with the entire team and work out the scope for every selected dimension. It doesn't have to be perfect and due to its iterative nature, the scope can be adjusted at the next iteration. But it is important to have the entire stream team understand what's in scope and out of scope for each dimension. You can reuse the aforementioned brainstorming exercises for scoping:

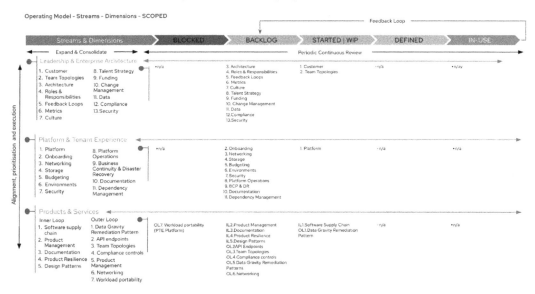

Figure 5.7 – All dimensions prioritized and scoped, with blockers identified and moved into WIP

The stream teams (are dream teams!) are now ready to deep dive into the different operating model dimensions.

Detailing your dimensions

With our streams set up and our dimensions selected, it is now time to get into the details of what the dimensions entail, how they guide us and how we reach the desired target states. Let's consider some suggestions regarding which stakeholder group could be the primary owner of an operating model dimension. But more than that, the descriptions for each dimension also provide different angles for each operating model dimension. They can be used to define content, boundaries to other dimensions, and cross-stream collaboration procedures on different aspects of the same dimension.

That said, there is not necessarily a right or wrong way to do this, meaning you can assign ownership to other streams and stakeholders. However, be aware that, ultimately, the stream that has the most relevant and timely information to make decisions is probably best set to own that dimension or the specific dimensions' aspects. This is because operating model dimensions rarely live in isolation; they affect each other.

Leadership and architecture

This dimension sets the tone for execution and empowerment and ultimately how much purpose and autonomy are felt and how much mastery can be achieved by the employed knowledge workers. However, it also indicates how teams are formed and how much trust can grow within a team, how Conway's law is implemented, and how budgets are created, to name a few.

The dimensions that you should consider for this stream are covered in the following subsections.

Customer

First things first: the very reason we do things. It can't get any more important and any closer to strategy and execution than the customer dimension. This dimension will define what CX the company wants to create – whether the CX is a differentiator or just a good-enough commodity to reduce CapEx. The CX can be used as a set of core metrics to determine success criteria. It helps decide what products and services are on offer, from where they are consumed, what acceptable wait times are, how many and how often new features are deployed. The customer dimension ultimately influences all other dimensions by providing guiding principles. These will trickle all the way down to your platform engineering teams and product teams, who will be defining their **service-level objectives (SLOs)** and **service-level indicators (SLIs)**. In some organizations, organizational strategy takes center stage when developing an operating model. However, with the ubiquitous digital mindset, this changes; it affects cloud and edge operating models because the customer moves closer and closer to technology-based experiences. There can be different types of customers, such as internal or external customers:

- The developer or "tenant" to the platform team
- The partner to the ecosystem team
- The partner's customer to the partner team
- The manufacturer to the control systems provider
- The retail end customer to the retail website developer
- The head of operations technology to the edge platform provider

Now let's look at the Architecture dimension next.

Architecture

This is probably one of the top three dimensions. Architecture decisions inform system and service design, as does organizational structure (Conway's law). We recommend that you include enterprise architecture capability models as an anchor model to overlay investments and maturity on the *as-it* to *to-be* roadmap. This then needs to inform the architecture dimensions in the other streams so that relevant decisions in the people, processes, and technology part of other dimensions are aligned. Capability models are models that show how mature specific organizational capabilities are implemented in your organization across people, processes, and technology.

Without being a bottleneck, your enterprise architecture capability should be established as a "design authority" that can veto decisions that go against the intent of your operating model. Your enterprise architecture will then be a core governance instrument for your operating model.

Team topologies

How are your teams set up, what is their purpose, and how are the teams interacting with other teams? These questions are what this dimension must answer.

The book *Team Topologies* recommends consolidating down to four different team types, such as a platform team, a stream-aligned team, a team for enabling other teams, and lastly a subject matter experts in a specialized knowledge domain, who are working on a, for example, insurance underwriting risk calculation module or training AI models as a "complicated subsystems team."

The corresponding three team interaction modes are X-as-a-Service, facilitation (helping), and collaboration. Let's say that the platform team provides the Core to Cloud to Edge **Platform-as-a-Service** (**PaaS**), while the DevSecOps team facilitates DevSecOps enablement across different product development teams. Finally, the data and edge teams collaborate closely to establish the best-fit patterns to collect, compute, filter, and enrich data at the edge.

Roles and responsibilities

To inform your headcount planning and employee development and talent strategies, we need to determine what roles we need and what they do. This is also informed by the enterprise architecture capability map. For example, if I have a DevOps team with an enablement charter, then I need to develop that DevOps skill and/or buy or rent it externally. Tools in this dimension use a skills matrix to determine what practice and training employees need and who and what to hire for. A hybrid approach can be to plan for anchor hires, who build up a team mostly from existing employees who are keen to learn new skills or move in from other parts of the business.

Feedback loops and communication

This includes communications plans, stakeholders, structure, tools, and update frequency across the platform and product teams, as well as internal and external customers. Fast feedback loops are vital for informing sprint planning, so this dimension is tasked with providing this capability while ensuring they've been set up and maintained.

Metrics and performance

Metrics and performance is another critical dimension. The right metrics don't need to just balance each other, like for example throughput VS stability. For strong team performance and to create and maintain a performance-orientated or generative culture, as defined by Westrum (refer to *Chapter 1*), we need team metrics that foster and encourage collaboration and knowledge sharing. At the same time, we need to be cognizant of the intrinsic motivators of knowledge workers, which are autonomy, mastery, and purpose (as per the author Dan Pink). These are sometimes also referred to using other, but similar, names, such as empowerment, competence, and relatedness. We mentioned the iterative nature of developing an operating model already. The advice here is to incorporate metrics and trends reviews, as well as ascertain the validity and appropriateness of certain metrics. We can do this by asking ourselves the following questions:

- Do those metrics balance other metrics?

- Are the metrics still relevant?

- What insight are we trying to gain?

People and culture

Research by John Westrum has shown that organizational culture is directly linked to organizational performance. This dimension determines how collaboration, shared risk, unhindered information flow, and novelty can be implemented as part of the organizational culture and hence can ultimately be observed through the actions and behaviors of the staff and contractors. Psychological safety is part of that. So-called "leaders" who hang people or teams out to dry or allow blaming when mistakes are made instead of pointing to what they have learned make it nearly impossible to establish a performance-oriented culture. Mistakes should be treated as opportunities to learn. For example, Google, under SRE practices, calls this "celebrate failures" as part of their postmortem culture, and people and teams that make the same mistakes repeatedly need help. This might demand lots of change in the executive layer. Who are the change agents that can make that happen in your organization?

At the same time, we need to consider the intrinsic motivators of knowledge workers. We need to consider salary and benefits and also from which $ figure upwards it makes no difference, whether people leave or stay. There is also a need to upskill, reskill, and cross-skill people or hire new talent (see the *Talent strategy and sourcing* section). There is a limit to team sizes and a suggestion that long-lasting teams provide the best ROI for organizations. Some suggestions state that cross-functional and autonomous teams are the bee's knees (meaning highly desirable), but failed attempts exist too. Above all, a continuous learning mindset is key.

How does your organization handle the constant and continuous cultural shifts through evolving and emerging practices such as DevOps, DevSecOps, GitOps, BizDevSecOps, FinOps, AIOps, MLOps, and EdgeOps? All this and more are reasons why the people and culture should be its own dimension.

Another cultural aspect is the creation of **Community of Practice (CoP)** groups within your organization that allow informal knowledge sharing across organizational silos to happen. This is a great way to foster organizational learning. In our experience, executive sponsorship (beyond free pizza and drinks) is required to keep those groups alive. Organizations can also make the CoP participation a corporate objective, assign metrics or OKRs around contributions, and include it in the **individual development plans** (**IDPs**) that align with career growth.

Capacity management

Capacity management is tightly related to metrics and can be combined with the metrics dimension. This is all about flow and establishing flow metrics, such as lead time or wait time versus process time.

At the same time, different work types and disruptions need to be tracked for improvement purposes. The most important thing you must do is distinguish work types in terms of planned work (for example, project work and routine maintenance) and unplanned work (outages and screaming VPs). If needed, you can further sub-categorize work types as your organization sees fit – for example, changes (arising from already completed work) and internal projects (software updates and server migration). Finally, you need to track "time thieves," as suggested in Dominica DeGrandis's book *Making Work Visible*. Let's look at these time thieves:

- **Too much work in progress (WIP)**: This is work that has been started but not finished

- **Unknown dependencies**: These are something that we are not aware of that needs to happen before we can finish

- **Unplanned work**: Interruptions that happen, preventing people from finishing their tasks

- **Conflicting priorities**: Tasks or projects that compete with each other

- **Neglected work**: Partially completed work that has been abandoned

Talent strategy and sourcing

Talent strategy is another important dimension that can severely hinder organizations if not connected to the strategy and the target cloud and edge operating model. Apart from cost-based headcount planning, the **talent acquisition** (**TA**) team needs to be experts in their respective fields and understand what and why they are hiring. More often than not, this is overlooked, but the TA team needs to have an intimate understanding of the company's technology strategy and operating model to scout for the right talent. Filtering out CV-on-paper superstars with little depth will save organizations many hours and dollars. Too often, many hours of key personnel are lost as they sit unproductively in interviews that should have never been scheduled in the first place. Another talent-related question is build versus buy. How much are we investing in staff training? Can we cross-skill from other departments? Do we offer graduate 3-year apprenticeships from 0-to-hero in data science, cloud, edge platform, or full-stack development? Who are we assigning mentors and putting on scheduled job rotations? What percentage of contractors should we hire at a maximum and why? Should we invest in industry superstars and make anchor hires who then build out a team around their core expertise? As you can see, this dimension is critical because it's also closely related to budgeting and funding and hence should sit in the leadership and architecture stream.

Enabling teams to continuously learn and develop the necessary skills is another aspect of this dimension. A good way to do this is to procure company-wide access to online learning and certification portals. This could include advising management to make valuable (not busy work!) certifications part of a bonus plan.

Your talent strategy can even serve as a control dimension for the technology teams making purchasing and architectural decisions. The skills you are hiring for usually convey if you are natively building into all hyperscalers and an additional multitude of edge platforms because it requires your TA team to hire for several teams with similar skills and salary expectations. Or are you abstracting and consolidating and re-using people, skills, automation playbooks, engineering best practices, tools, and accumulated knowledge? If so, your TA team can post job ads while looking for the relevant skills that support this strategy.

Service-level agreements, business continuity planning (BCP), and disaster recovery

This dimension in the leadership and architecture stream establishes a framework for other streams to lean on. CX-based **service-level agreements (SLAs)** inform workloads and hence the product and services team, which, in turn, relies on the platform team to support the required SLAs and, in the case of catastrophic failure, BCP and disaster recovery efforts. **Recovery time objectives (RTOs)** and **recovery point objectives (RPOs)** should be used to establish well-defined SLAs. RTO and RPO are specific recovery objectives that determine the time and data point requirements for restoring business processes and data. These objectives are typically documented in SLAs to establish mutual expectations between service providers and customers regarding recovery time and data loss. These SLAs serve as a contractual framework for monitoring and evaluating the service provider's performance in meeting the agreed-upon recovery objectives. If an organization wants to make service (instead of site) reliability engineering part of their distributed operating model, then this dimension in the leadership and architecture stream needs to provide the foundation. This can be done by establishing, at first, SLAs for each workload across core (on-premise), cloud, and edge, and requires the definition of SLOs to be the benchmark. This benchmark is a range between the best possible and least acceptable. While best possible is an engineering decision concerning how much effort (read: money) is spent to get as close as possible to 100% availability and zero latency, least acceptable is driven by still-good-enough CX. The platform and product teams can then look at the required SLIs to help ensure that the SLOs are met and CX is not impacted. Metrics to look out for here include request latency, throughput (that is, traffic) errors, and saturation. These are also referred to as the Golden Signals.

To stay below the maximum outage times in compliance with the contractual SLAs within a distributed or edge scenario, organizations might need a different approach. For example, instead of backup/recovery for a monolithic core system, in a distributed context, SLAs might be adhered to through resilient designs such as self-healing, event-driven asynchronous services providing eventual consistency, active-active workload and infrastructure deployments or other types of redundancy.

Branding and change management

It might sound a bit counter-intuitive or even silly, but I've seen it with my own eyes. Teams who start by breaking new grounds – be it building an entirely new digital bank, deploying containerized virtual network functions, designing microservices, building or moving to an on-demand cloud, or building out edge environments – are sometimes exposed to demeaning behavior by the "outer circle" or change-resistant folk. It's valuable to understand that those behaviors are not personal but often fear-based and need to be addressed holistically, not on an individual basis. Regular update sessions or meetups with pizza and beer showcasing the latest developments and sharing learnings are a good start. Teams can be branded, for example, as digital ninjas with stickers on laptops or T-shirts. And when the first successes set in, those laughs will quieten and people will knock on the door, asking if they can join or learn more. Branding also helps create better awareness and affinity as more and more internal people are initially curious, then observe, then participate, and eventually champion the initiative. We suggest that you take a close look at branding an initiative to maximize its reach.

Funding and budgeting

This requires a good understanding of the other dimensions, which serve as input to funding and budgeting. Historically, most organizations set limits on investments and budgeting, as well as project or program funding, based on annual, 3-yearly, or 5-yearly budgeting cycles. While a program-level initiative might get multi-year funding, it is important to establish flexibility inside those funding constraints. For example, product managers need to be empowered to shift gears or change directions if market demand changes. This often requires budgetary controls to change and recognize that Agile delivery methodologies do not deliver against a fixed set of outcomes.

Procurement and vendor management is another angle that affects budgeting, as is internal micro-control. A type of issue in this dimension is micro-level budget control. It can become (and dare I say, usually always is) a massive roadblock if you need to get approvals from several levels above in order to shift spending from your compute to storage, for example. That's why a shift from project to product funding can be vital. The product manager should have full control of the product budget range.

Data

Because data is the new fuel, data security and data access need to be balanced. Organizations can't make it too hard to get access to data as it stifles progress. At the same time, data – especially sensitive data – needs to be kept secure, and specific data compliance frameworks must be adhered to. Data access policies need to be defined and centrally governed. This can be achieved via central data access policy management combined with distributed policy enforcement mechanisms. And your operating model needs to enable this. Allowing organic ungoverned data storage locations or data access is discouraged in a distributed environment. It will most likely get out of hand and not yield the desired results. We do not recommend a centralized **master data management** (**MDM**) approach. Previous initiatives we worked on in different organizations could not make a centralized master data management approach work. If you decide that your operating model needs to have data entities defined and data stewards or data product managers assigned, then the recommendation is to employ a **domain-driven design (DDD)** approach instead. This will allow different parts of the business to own core data attributes and define the best-fit data access policies with a smaller stakeholder group, which aids decision-making.

When you hear that data is the new oil in the digital economy, you might want to differentiate more and consider data as the new crude oil, with information (data with semantics and context) as the refined oil and insights as the engine fuel. But we do believe that data needs its own operating model dimension, given the many failed MDM initiatives, data quality issues, and the inherent data gravity pull. Thinking about how to define and serve your data, manage data policies and associated controls, and evolve data into information and insights while allowing access in a distributed context is a good starting point.

In a distributed environment, it is important to have a robust data classification model and entity register to understand where in the operating model data resides, including data lineage. This informs the controls and SLAs, which need to be in place to secure and assure the information. When onboarding information assets, vendors, or partners, it is important to link data entities, which will be contained with the information asset or serviced by the vendor to gain an understanding of potential breach vectors or areas of sensitivity within the operating model. This should also be expanded to services so that an organization knows which services expose the organization to the greatest risk. This shows that security is a cross-cutting concern across many different operating model dimensions and streams.

Distributed security and compliance

Even though "distributed" is intrinsic to all dimensions in a hybrid cloud and edge operating model, it deserves a special mention in the security and compliance dimension. We will dive into specific security topics such as container runtime security in the other streams, but this dimension in the leadership and architecture stream needs to set the direction on things such as supply chain security, non-repudiation, policy-driven data access at the field level, and infrastructure components for a zero-trust architecture.

Think about how to raise the tide within your organization. "*A rising tide raises all ships*," the saying goes; the same is true for security leadership. Nothing is ever idle in this dimension. Communicate what security and compliance control frameworks your organization implements and why. What industry compliance frameworks exist? What security frameworks and best practices do you rely on? Be clear about why Essential8, STIG, NIST, PCI-DSS, supply chain security, or post-quantum cryptography is or is not fit for your organization.

Logging is another fundamental aspect of an operating model since its potential home is in security. A guiding principle of "*All logs must be shipped to our central logging instance*" makes sense in a distributed environment. This not only allows for anomaly detection but also establishes observability and tracing of transactions across multiple geographically distributed environments and system components. All technology components mentioned across all operating model dimensions should follow these guiding principles.

External factors

External factors are another key aspect of your operating model. Therefore, it is recommended that you have a dedicated focus on understanding these external factors and that you track them accordingly.

Different from external dependencies (which are established and known), external forces can be either positive, neutral, or negative; known or unknown; and they may be controllable, influenceable, or uncontrollable.

Examples of external forces include economic conditions, technological advances, competition, regulatory and legal changes, social and cultural trends, mergers and acquisitions, and political developments. These factors can affect the demand for products or services, influence consumer behavior, shape the business environment, and impact the supply chain.

That said, a change in strategy is not part of the operating model. The operating model has to support the strategic execution but not define it. While changing demand and change in consumer behavior is something the strategy or strategic goals need to be adjusted to, the operating model can make provisions for some of the more foreseeable changes, such as legislative changes (neutral and uncontrollable) or mergers and acquisitions (positive, controllable, or at least influenceable).

In your operating model, you can categorize potential changes in the aforementioned areas as risks and come up with adequate risk mitigation plans.

For example, when I was working in one of the largest banks in the APAC region supporting their super-regional strategy, in our next-generation enterprise integration architecture, we catered for **merger and acquisitions (M&A)** support through business event subscriptions. This allowed newly integrated business units to subscribe to relevant business events (for example, payments or loan origination) from other domains. This asynchronous approach made it much more feasible than the originally designed synchronous API mechanism.

Here are some good sources you can regularly ingest to help you keep abreast of changing external factors:

- Your organization's strategy documentation
- Analyst papers on emerging short and long-term trends
- Updates from industry regulators
- Professional networks
- Industry events

Success criteria

Now that we're done with our dimensions, I'd like to spend a moment on *What does success look like?*

When you establish success criteria within your operating model, there are a few things to consider:

- Manage outcomes versus activities. Set OKRs to align with what you want to achieve and create accountability across teams. Historically, there was a focus on busy work – that is, tracking activities resulting in different OKRs for different teams – which is why value creation can be missed. While these sorts of metrics might have a place somewhere, your operating model needs to be measured against the business value that's created. This in itself can be an onerous task, but asking the relevant questions of your business stakeholders is a great way to track the progress of your operating model, especially as time, effort, and costs accrue.

- Build goals from the top down and bottom up. Operating model dimensions that overlap need to communicate in both directions so that nothing can be missed or, worse, contradict. This helps you ensure that goals align across streams and dimensions.

- While everyone talks of "transformation," the magnitude and complexity of building your distributed operating model require careful orchestration and coordination. It must reflect the greater scope of the effort. As mentioned previously, this is why stakeholder management is important across the entire organization – for instance, working closely with HR to hire the right talent, collaborating with developers and business sponsors to deliver outcomes, and bringing in people with sufficient domain expertise to manage complex decisions.

- Prototype – build, measure, learn. One of the things organizations learned early on while creating their operating models was that assumptions and designs did not always meet reality. So, while they were designing their models, they created prototypes to build skill and awareness within the team of what works and what does not. We recommend layering prototypes in all of these dimensional items to validate assumptions and decisions within the model as soon as possible. This will reduce the risk of "ivory tower designs" and overspending through overengineered not- fit-for-purpose approaches. Use those prototypes to learn what metrics matter to achieve quality outcomes.

- Build foundational guiding principles to enable quick decision-making. For example, discuss whether the operating model will be implemented as an automated as-code model, such as Terraform or Ansible, or as a manual CLI or ClickOps model. Getting these fundamentals addressed across the distributed estate at the beginning will also drive what software and applications will be contained within the model and help you gain insight into what metrics count.

Leadership and architecture stream summary

With that, we have covered a comprehensive but non-exhaustive list of operating model dimensions and aspects for the leadership and architecture stream. You might find others now or later or just go with what we've covered here – all of which is a great start.

Most of these dimensions are connected or set the context of dimensions in the subsequent streams, which we will explore shortly.

As seen with funding and budgeting and how tightly those dimensions can be inter-connected, it is important to recognize the iterative nature of developing an operating model. We believe that this is the only way. It recognizes the fluidity that new technologies, changing external factors, and continuous learning brings daily.

The operating model's documentation needs to be made available via an always-accessible online channel. This can be a wiki or a Git repository, for example. It aids in the information flow and is needed so that it can act as input for workshops for other streams. However, the iterative and interconnected nature of the operating model dimensions also requires formal knowledge sharing and cross-stream synchronization. In the beginning, this needs to happen quite frequently and can then be spaced out as the operating model matures through operational feedback.

We believe that the features of a successful operating model are that it's built by internal community with open culture principles, meaning stakeholders have had their say, it's fully documented and accessible, it has clear roles and responsibilities, there is active participation by all stakeholders across all streams, the dimensions have been jointly agreed upon and are reviewed and improved on an ongoing basis, all stakeholder groups are represented, and that metrics are captured and analyzed for progressive improvement.

The way your organization develops your distributed technology operating model can become a great blueprint and leading example of how to build a generative, performance-oriented culture.

Now, let's dive into the next stream and explore the operating model dimensions there to get you set up for your distributed future.

Platform and tenant experience

Ownership lies often with a central team such as platform engineering (sometimes called DevOps engineering or just Platform Team). The job at hand is to provide multi-tenant application infrastructure to their LOB **application development (app dev)** teams, inclusive of appropriate governance, policy, security, and overall guardrails.

The definition of **cloud platform** can range from infrastructure-only components such as compute, storage, and networking to "whole of (public) cloud" approved proprietary services, which often manifest themselves in hundreds of disparate services being coupled together or ready-made open source platforms such as Red Hat OpenShift.

This dimension is about preventing "we'll build it and they will come," More positively formulated, it's about creating a great user experience or CX for the cloud and edge users. Typically, these are developers or COTS application owners who want to develop, test, continuously integrate, and deploy their workloads. To achieve this, you need to start by defining who your customers are and what they care about.

The dimensions for the platform and tenant experience stream you should consider are covered in the following subsections.

Platform

At first glance, this dimension might look like *Captain Obvious Strikes Again*, but in reality, it can be quite hard to do: we need to define what we mean by platform or, more specifically, what our hybrid multicloud and edge platform experience should entail.

A platform is the *undifferentiated heavy lifting* you are abstracting away for your tenants to help enable your product teams to drive an outstanding and differentiated CX across your products and services. This dimension includes tools and technology, shared responsibility considerations and platform operations, workload SLAs, and how to achieve them:

- Do you have ARM, x86, System Z, or Power processor architecture-dependent workloads?

- Do you have teams, services, workloads, or data across multiple public hyperscalers or colocation data centers?

- Do you prefer managed services or SaaS platforms?

- Do you need a DPU, FPGAs, or GPUs for model training and inference at the edge or for AI/ML?

- Do you have **high-performance computing (HPC)** or **hyper-converged infrastructure (HCI)** requirements?

- Do you rely on specific hardware vendors for a **hardware security module (HSM)** or **Single Root I/O Virtualization (SR-IOV)**?

All these questions play an important role and should be answered as early as possible because they impact your tenant experience, and hence adoption, and what options need to be enabled and catered for in other dimensions.

It is also recommended that you understand the business goals and outcomes the organization is focusing on in the next 1 to 3 years. This will give you additional insight that can be added to the platform's vision and architecture as you build your operating model out. For example, if your executives often talk about how beneficial AI would be to the business AI or have a pet project in the quantum realm that has not materialized yet, then you can bake those considerations in:

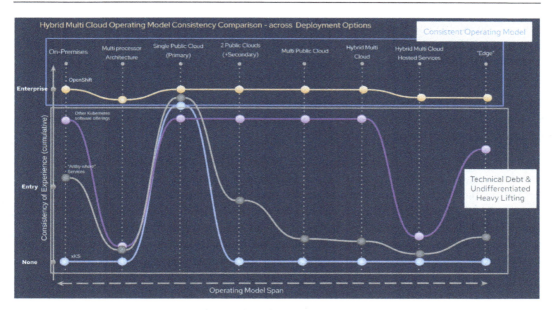

Figure 5.8 – Comparison of platforms and services while depicting
technical debt and undifferentiated heavy lifting

The dots in the preceding figure indicate different maturity levels in terms of enterprise-readiness – that is, are the possible choices mature enough for your needs? Mature can mean secure, supported, training and documentation is available, it's patched by vendors with security response teams, and it's actively developed further, ideally with a strong open source community behind it, among other things. Here, it's important to have a clear view of why you are going with option A instead of option B and then document it.

When I had to make product selections in the past for enterprise integration platforms, I created a weighted list of features and functionality. The features listed are the core critical features, while the weight indicates how important that feature is and hence how much it impacts the decision-making process. This also helps you document your decision-making (trust me, someone will ask someday why this decision was made). For example, as per the question at the beginning of this section, if you have workloads on ARM, System Z, Power, and x86 and want to reduce operational burden by using first-order services in hyperscalers, plus you have plans to utilize edge computing, then a software-based enterprise-ready platform can be the best choice. If you are a start-up and your main goal is to prove that there is a market for your product idea within the first 12 months, then perhaps a limited amount of services from a single hyperscaler is the better choice. This is where your weighted list of needed features and functions concerning your target platform vision helps you select the best fit for you.

If you don't break out metrics as a separate dimension in this stream, we recommend that you measure business value. This can be the number of applications onboarded, time to revenue per app, platform performance, and adoption rate. As we mentioned earlier, just find balancing counter metrics to make sure you find your equilibrium and deter attempts to rig your metrics.

Digital sovereignty and cloud reversibility (for example, repatriation to on-premises infrastructure) are additional angles that warrant consideration regarding what your operating model for distributed workloads looks like in the platform dimension.

Digital sovereignty aims to improve cyber resilience, protect economic interests, and safeguard sensitive information, as well as protect national interests. It refers to an organization's ability to exercise control over its own digital data, networking, and general technology infrastructure. This can be done by enforcing controls, as required by law, and defining the exact flow of information, including crossing into different jurisdictions. The overall aim is to reduce dependencies on a specific technology provider, be it in-country or foreign. Digital sovereignty is built on core principles such as public cloud reversibility onto a fully controlled hybrid cloud and edge environment, a trusted software supply chain, as well as data federation, including computational governance and data as code. It becomes applicable for situations where your cloud provider, for example, is under US jurisdiction and your organization is a non-US-based organization. This example is just as applicable to any other combination of countries with incompatible privacy laws and data protection mechanisms.

Conversations about where your data lives and where your infrastructure sits should be made a priority. And that is not just a "where" in terms of geolocation but also what jurisdiction the internet router that your traffic passes through is under. Be sure to check those details in your contract or SLA.

The US isn't the only country with data surveillance laws. The United Kingdom has passed the Investigatory Powers Act, giving security services in the UK permission to use a wide range of tools for surveillance and hacking. Knowing the laws of the country your data resides in is paramount to protecting your organization's and client's data. This is effectively non-compliant with the European **General Data Protection Regulation** (**GDPR**). Since 2020, multiple relevant court activities have been covered under the term **Schrems**, named after the Austrian activist starting those efforts. This is an evolving space, so we believe this warrants investigation when you start creating your operating model. But never assume you know where your data travels and who owns that infrastructure you use. You can do the following test:

- VPN into a country in Europe (e.g., Germany)
- Select a major enterprise and find their website URL (`Allianz.de`)
- Traceroute to that URL (`traceroute allianz.de`)
- Lookup all the IP addresses via an IP location service (e.g., `52.93.42.138`)
  - Check geolocation (e.g., United States)
  - Check who owns the network infrastructure (e.g., Amazon Technologies)

Figure 5.9 – Traffic originating in Germany destined for a German website
travels to the US and through US-owned infrastructure

Does your business benefit from cloud reversibility? That is, is your data, operations, and infrastructure layer abstracted so that it is possible to rebuild your entire application stack, including data, compute, and networking, in a different location via automation? This must include your engineering and release and deployment processes. Hence, your tooling needs to support hybrid multicloud and edge locations, independent of the technology provider. Some customers of ours regularly rebuild their tech stack every 60 to 90 days. This approach can also be reutilized if cybersecurity breaches occur or intrusions are detected as it allows for rebuilding, and increasing monitoring efforts while investigating what happened.

If confidential computing is part of your workloads, your operating model will have provision for that, too.

If you build your platform on managed services and even if those managed services are "borrowed" from the open source community, do you know what the additional operational source code, processes, and data look like? Can that managed search service still be regarded as open source?

All in all, there are many different viewpoints that organizations can look at within your platform dimension.

To conclude, this dimension's concerns should cover disconnected platform operations, isolation requirements and the available multi- and single-tenancy models, environments across virtualized, bare-metal, on-premises, and third-party managed implementations, workload requirements for HPC, low latency, GPU, TPU, and other specialized workloads, environment specifications (ideally as code), operations only environments to test emerging procedures (in an SRE context, for example), build quality control for environments, including necessary verification testing, environment profiles supporting workload categorization, and associated requirements, control plane endpoints, cost management, and the control plane.

Control plane

Another often overlooked but vital aspect of your *core to cloud to edge* platform is the control plane. We borrowed and extended that term from networking, but as you can imagine, in a distributed cloud operating model, the control plane deals with more heterogeneity and, as a result, is much more complex than representing merely a network topology through routing tables that define what to do with incoming packets.

Often referred to as a single pane of glass (and hopefully not a glass of pain), manageability, observability, tracing, monitoring, and alerting are vital for knowing what's going on in your environment.

The control plane needs to be catered for from day 0, which means the control plane needs to be set up when all the other system, software, and infrastructure components are set up and configured before they are deployed to production.

The first test for your control plane is to monitor and aid day 1 operations. Day 1 operations refer to the tasks and activities that occur when new systems, software, and infrastructure components get deployed to production and are made available to your users and customers for the first time.

From day 2 and beyond, your control plane dashboards need to support the ongoing tasks and activities that are necessary to maintain and support your environments. These operations typically settle in after the first day of production use and are hence referred to as day 2 operations. Their main aim is to help ensure that your environments function properly and grow with the changing needs over time.

It's never too late, but if you haven't purposely designed your control plane and you are adding edge computing on top of your cloud initiatives, then you are behind. Organic growth in this dimension is not what we want.

If you don't have that single place where you can get a consolidated view across your distributed cloud and edge estate, be it log files for debugging, transaction tracing, or business events, then the risk is that you will be flying blind into your distributed future.

While it's relatively easy to use your built-in hyperscalers cost and events dashboard, it requires more work in a distributed cloud scenario, with different log file formats, cloud services, infrastructure components, configuration approaches, and so on. It's worth spending some spikes to evaluate the most appropriate processes, tooling, and abstraction layers.

Advanced control planes our customers have built allow them to automatically create a new branch in their source code repository while automatically building a production environment replica, including relevant production data for debugging and bug-fixing purposes or creating a beta environment for their beta customers. If you reach that level of control plane sophistication, you deserve half a day off.

In an Edge context, further control plane considerations are necessary. Edge device provisioning includes either manned or unmanned operations, or a mixture of both. Depending on the scenarios different approaches and technologies across secure or unsecured networks for remote hardware and boot attestation, updates, rollbacks, management and decommissioning or removal of network access need to be chosen. End-to-end Edge fleet management requires a careful definition of the context, such as the type of data, the functionality, and the hostility your devices will be exposed to.

Another growing aspect is sustainability concerning power consumption and the sustainable use of resources. If you look at new GPU families being released by chip manufacturers, you will see that they still list the performance benchmark. However, apart from those numbers, they are not shy to tell the customers that their chips consume less power than the competition. So, if sustainability is big in your industry and organization, you might even want to break this out as a separate dimension. If it's not, then you can still lead the way on that topic with your operating model.

Architecture

Informed by the enterprise architecture's business capability model and corresponding roadmap, the platform architecture takes on the supporting capabilities to build (SRE or GitOps, for example). These supporting capabilities are established when people, processes, and technology are established and the platform architecture and tooling align across on-premises, the cloud, and your edge environment. More importantly, however, the platform architecture needs to decide to what degree low-level details such as compute, storage, networking, operating system, platform life cycle management, and platform security should be abstracted away for the platform users. This includes low-level details such as provisioning and securing bare-metal environments at the edge, how CI/CD flows are connected with the inner and outer development loop across different environments such as dev, test, QA, UAT, pre-production, and production, and what observability metrics and monitoring you put in place to assist a full cloud governance and compliance approach.

Good concepts to familiarize yourself with while working out your distributed cloud and edge architecture are the CAP theorem and the fallacies of distributed computing.

The CAP theorem states that no distributed system is safe from network failures – called network partitioning. In the presence of a partition, the architecture needs to trade off consistency and availability. When choosing consistency over availability, the system will return an error or time out if particular transactions cannot be processed due to network partitioning. One way to handle this is to ask the user to retry. When choosing availability over consistency, the system will always try to process the transaction or return the most recent available version of the information, even if it cannot guarantee it is up to date due to network partitioning. This can be achieved through asynchronous processing or caching, for example.

Again, those architectural decisions will not only impact how the workloads (the platform tenants) are architected but also what tools and processes are supported by your platform engineering team.

The other concept I mentioned is called the fallacies of distributed computing. It consists of the following aspects:

- **The network is reliable**: We need to architect, design, and code for network failures

- **Zero latency**: Consider delays in transaction processing, query results, and the impact it has on CX

- **Bandwidth is infinite**: Beware of bottlenecks

- **The network is secure**: Consider what the zero-trust architecture brings in as input into your architecture

- **The topology doesn't change**: Hybrid multicloud and edge almost certainly have different network topologies from day 1 and new developments such as WIFI6 or 5G/6G and beyond have already announced that change

- **There is only one administrator**: Sovereign cloud efforts and network automation look like they dampen the impact of this fallacy, but just considering that edge computing is likely to have **operational technology (OT)** instead of IT as the main stakeholder makes this an important aspect to consider

- **No transport cost**: This is not only about hyperscaler data egress costs that have often been "forgotten" but also network architecture changes and the maintenance and change management thereof, serialization/deserialization overhead, protocol overhead, network latency, bandwidth limitations, and the processing power required for data encoding and decoding

- **The network is homogeneous**: Again, as stated previously, a distributed future with hybrid multicloud and edge workloads cannot assume a homogeneous network

While these topics are highly relevant for workloads and applications, the provisioning, operations, ongoing monitoring, alerting, debugging, and resulting remediation, as well as the definition of the SLAs and SLOs and how the SLIs of the platform itself are captured, make them highly relevant for the platform architecture and platform engineering.

Data

Lots has changed in the data dimension: from flat files to relational databases to data streams to NoSQL databases to data warehouses, data lakes, data lake houses, BigQuery, and distributed data mesh architectures. Here, **artificial intelligence (AI)** is a consumer with a thirst for more and more data. If what is real time?, then it's even better as that means that we will have to leave our thinking caps on for a while. Data is the new oil, gold, or fuel we have heard about multiple times. Data volumes are also bound to grow exponentially. But what's most important is to recognize that while data is your asset, it's also the biggest weapon to hold you ransom via storage and data egress costs. You could say that data is an asset but if it's managed incorrectly, it is a liability at the same time. If left to grow organically, it will become even more of a liability, with growing costs, reduced value, and, ultimately, becoming unsustainable. Cleverly introduced differences in data storage pricing keep you busy thinking about cost control when the root of the problem is that every time you want to access or move your data, you might have to pay for it. So, the question is, how can we be clever about data in an inherently distributed hybrid cloud and edge environment? A data mesh approach is best suited for a federated and distributed environment consisting of hybrid cloud, multicloud, and/or edge locations. We need to aim to not duplicate or physically move data, as well as leverage all the existing data jungles (warehouses) and data swamps (lakes), and lake houses. But like with any good operating model dimension, there are other considerations as well.

There is lots of open source data tooling out there. First, there's software-defined storage. CapEx in a long-term context such as data storage is a valid option. Relatively cheap open source software storage solutions are also worth exploring, especially if you consider that most established colocation vendors have big overseas connected internet pipes and direct connections to the hyperscaler cloud providers. This may make you want to consider making your co-location the master data location while the next-gen killer app sits on a public cloud or at the edge. You need to explore if it is cost-effective to store data in a public cloud and pay for data to be transferred between your data center, clouds, and edge locations in a more and more distributed future.

On the data processing plane, there are open source memory data grids, messaging infrastructure, relational and non-relational databases, Kafka implementations, data provenance, and data access policy as code frameworks, as well as distributed query engines. Given that those tools run in most locations, they need to be considered and probably even favored over proprietary location-specific implementations that require different skill sets per location and hence different operating procedures per location.

Apart from technology, there are also data modeling aspects that contribute to how the platform provides data. Please refer to the data dimension of the leadership and architecture stream for more details.

I think it's fair to say that getting the data dimension of our operating model right yields not just one of the biggest ROIs but also top and bottom-line revenue and profit contributions.

We will dive deeper into general security aspects in the security dimension. But for data, the standards for data in transit and at rest remain. Confidential computing is extending this dimension just like expanding on **zero-trust architectures** (**ZTAs**), where we need data field-level access control and a supporting real-time access policy-based engine. That engine controls where the data can be accessed from. Access is granted or denied based on where the data and the requestor resides. These topics are related to the Sovereign cloud context as well, which is beyond the scope of this book.

Within a distributed environment, the scalability and efficacy of your data dimension depend on two related functions:

- Transferring data between your data center, cloud, and edge locations in an enterprise-grade, cost effective, and timely manner
- Making the production and consumption of data for tenants as easy as possible and as secure as required

When adopting the necessary federated and distributed architecture paradigm, the platform team needs to work with the tenants on how and if they are moving to a more service-oriented data products architecture. This setup allows organizations to share data in a structured and governed way, as well as facilitate the interaction between data services and data products through well-architected APIs:

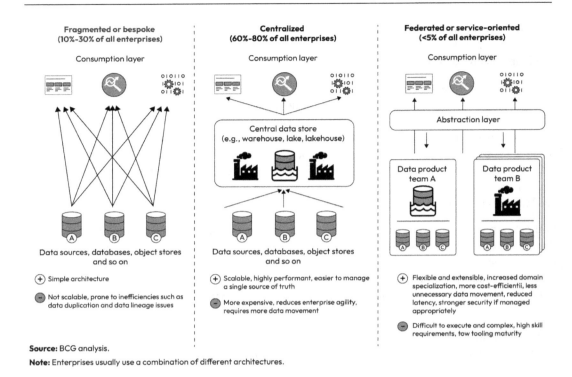

Source: BCG analysis.

Note: Enterprises usually use a combination of different architectures.

Figure 5.10 – Different data architectures based on enterprise maturity

In a distributed future, the BCG analysis shown in *Figure 5.9* is to be overlaid with different locations: a fragmented data architecture will most likely become a bottleneck or at least inefficient and costly. Centralized data store locations need to be assessed concerning their impact on how timely data can be made available and then, subsequently, accessed. Finally, a service-oriented data architecture forces additional complexity upon enterprises that opt for federated and distributed abstraction layers. That said, and given the immense data growth and real-time requirements we are seeing from edge computing use cases alone, a federated data architecture might be the only feasible approach for a scalable distributed operating model.

All these aspects need to be considered in the data dimension of your operating model. Because we are currently in the platform and tenant experience stream, the data dimension needs strong involvement from our tenants – that is, the workloads and applications or products and services teams, as well as the leadership and architecture stream. This is because a federated data mesh-like approach needs strong computational governance, traceable data provenance, versioning for point-in-time audits, and field-level access control, which, in turn, overlaps with the ZTA.

Security

As the saying goes, "*Security is a process, not a product.*" The platform and tenant experience stream provides guardrails and guidelines via tooling, automated processes, and documentation/training around security and compliance for the tenants. Tenants are the workloads on the platform. On top of that, this stream is responsible for the security of the cloud platform itself as part of the operating model definition.

It might sound a bit counter-intuitive, but the trick is to reduce security-related cognitive load for your tenants so that they can focus on what they need to get done, which can be things such as deploying a new version of your COTS suite, applying patches, developing a new feature or fixing a bug. The easier the life of your tenants onboarding and operating workloads is, the more adoption and love your cloud platform will get. We often hear about "shift security left," which is basically about moving security concerns earlier in the development life cycle. But putting just everything on the plate of developers is certainly the wrong idea. And while the cloud platform probably can't provide all the security tooling and security automation for all tenants out of the box, the operating model should provide a fit-for-purpose security posture baseline and enforce deployment-preventing policies around it for workloads which don't meet that baseline.

To summarize, the security dimension in this stream has two branches:

- Your distributed cloud platform, which needs to be built from enterprise-grade secure and supported components that are backed by a financially stable supplier who provides training, documentation, and a product security response team. Your supplier should have internally established software supply chain security practices, as well as proven experience with working on embargoed vulnerabilities.

- Security enhancing tooling for the tenants to select from (for example, automation, CI/CD, or certified container images).

The compliant and in-organization approved-to-use software components from your suppliers need to come with a traceable provenance via a **software bill of material** (**SBOM**) for specific versions and patch levels. This should apply to all your proprietary and open source software components, such as operating systems, databases, DevSecOps tooling, messaging infrastructure, storage and networking, and proprietary hyperscaler services. A rule of thumb could be that the more heterogeneous your environments (on-premises, public cloud, edge, and different processor architectures) are, the more likely is it that software components are a better choice than proprietary services because skills, processes, and tooling are reusable, whereas knowledge of proprietary services that are only available in one location are hardly reusable. This affects your efforts to maintain a consistent security posture from core to edge to cloud. Tooling that can consistently execute your security controls across all your platforms should be preferred. This starts with simple cross-cloud/platform automation up to specialized security tooling for field-level data access security, runtime security, intrusion detection, and so on.

Some organizations opt to execute a fully automated rebuild of their cloud environments every 60 or 90 days. Apart from keeping configuration drift in check (which can be a security concern in itself), if your organization opts to build this capability, then this can be used as remediation for security breaches and intrusions. Automated certificate rotation is a sub-component of this.

Further security features and tooling to remove cognitive load for your platform tenants might be storage encryption, data transport encryption in transit (mTLS), PKI, certificate-based authentication, secret management, code linting tools, container-based IDEs, test automation frameworks, software supply chain assessment tools for libraries, and container images.

Security of the cloud considerations includes mTLS, certificate management, key/certificate rotation, image scanning and signing, hashing, booting only off non-compromised hardware, tcp dumps, intrusion detection, monitoring, logging and alerting, tooling for load, throughput, and latency measurement, and automatically enforceable security policies.

If your platform and cloud service provider is in the business of offering open source software as a service, then it's worthwhile enquiring about what the added operational IP is and what third-party libraries and components are used to operationalize that software.

In summary, the security dimension should consider encryption, key rotation, software supply chain security, secrets management, security benchmarking, compliance control automation, certificate handling, transport layer encryption and termination points, audit reports, user management and roles, penetration testing, security breach remediation, service account management, least privilege access enforcement, two-factor authentication, SIEM integration, operating system security mechanisms such as seccomp, control groups, and namespaces, identity management, role-based access control, how ZTAs are implemented, and finally, compliance and security control implementation from technical areas such as STIG, NIST, CIS, and Essential 8 to industry frameworks such as PCI-DSS and HIPAA.

Tooling, automation, and self-service

This dimension can be considered orthogonal as it can touch many of the other dimensions in the platform, including the tenant experience stream and beyond. Part of this operating model dimension is to think about what self-service and automation can be introduced to remove bottlenecks and dependencies and reduce cognitive load. As mentioned previously, you can decide whether a type of automation is even a foundational guiding principle.

The term **tooling** can be applied to different levels of sophistication. It can be sophisticated, such as a comprehensive **internal developer portal** (IDP) that helps onboard development teams and make it easy to transition workloads between the inner and outer loop. Simpler tooling or toolchains can be a specific source code repository, container image registry, out-of-the-box CI/CD tooling or code linting plugins, and pre-provisioned compliance and security enforcing deployment and delivery pipelines. The sophistication of the tooling and the resulting dependencies also impacts resourcing and team structure. The more that teams depend on specific automation components and tooling, the more important it becomes to have dedicated people to underpin the supporting platform team. The more locations that are supported, the more the supporting platform team needs a roster and alerting to not increase wait, lead, and cycle time as it will hamper flow and time to business value.

The tooling, automation, and self-service dimension of your operating model determines the resulting platform interactions. Interaction can vary on a scale from being manual to being various degrees of automation to the target cloud environment and the location being invisible. The choices you make here will impact other dimensions and their roadmaps, from their starting points through transition states to the ever-elusive target state (To-Be). This then impacts the platform-supplied tooling and processes.

If you have difficulties conquering the vast space of automation, you can think of it in two different ways:

- Help reduce the cognitive load of your tenants. With this in mind, you have a good starting point, even if your vision is to establish a real **SRE** (cloud platform **service** (instead of **site**) capability.

- The SRE capability itself. The SRE aims to dedicate time to engineering tasks that ultimately automate your end-to-end environment provisioning, including **quality assurance** (**QA**) efforts, stability, availability, performance, compliance controls, and the security of the platform or environment itself (as opposed to the workloads).

Additional topics to consider in this dimension are environment creation, automation in support for disconnected environments, asset and source code repositories, API endpoints, pre/post environment creation automation, pre-/post-workload deployment tasks, CI/CD tooling, different tooling deployments per region, availability zones, clusters or namespaces, a policy as code setup, namespace management, certificate creation and rotation, LDAP integration, user and service account creation, storage class creation, including thin/thick storage provisioning, SSO integration, integration in your change or asset registry, secrets management, programming language support, backup and restoration testing, workload quote rules, application performance management, monitoring, alerting, tracing, logging and observability, upgrade and rollback automation, environment scaling, metering integration, AIOps model training and deployment, versioning and labeling, real-time error budgets, and SLO monitoring.

In summary, the best areas for automation are cloud and edge platform or generally environment deployments, including pre and post-deployment operations that cover day 0 to day 2 concerns and deploying and setting up infrastructure in terms of networking, storage, and workload deployment, including dependencies.

Environments – VPCs, regions, availability zones, locations, and clusters

This operating model dimension gives you pointers to consider when defining the environments in your distributed cloud operating model.

When we say *platform team* in an environment context, we mean the team that looks after the different environment configurations, regardless of whether it's VPCs, regions, availability zones, or clusters.

I know some people think you only do proper DevOps if you test in production, and it makes for a good headline, but realistically, you don't want to send an email to your 1,000 tenants and paying customers with the subject: "Ooopsie" or "Production environment will be available again in 3 hours." You also don't want to explain to your CEO how you caused a revenue drop of 4 million dollars because you believe real DevOps practitioners test only in production. Or at least I don't. So, in terms of environments, it makes sense to come up with a list of environments, their purpose, and who owns/manages/operates/updates and maintains them. You can build a RASCI around that if you want to be thorough.

Here are some possible environments you can consider:

- **Operations**: Mimics production and is for operations to test new features, functions, services, or updates and rollbacks.

- **Development**: In engineering and software development organizations (and which organization wouldn't be nowadays?), treat it like your production environment to keep your development teams creating business value. Even though developers could be productive on your inner loop alone, only code in production can create business value.

- **QA/integration/user acceptance**: A test environment with near production-like functionality.

- **Performance**: If performance, latency, and response times matter to your customers, then this test environment needs to be set up with production-like compute, network, storage, data, and applications to help you examine and test the CX in realistic load scenarios.

- **Pre-production (also called staging)**: Some organizations do UAT, performance testing, security testing, and final pre-production validation in a single pre-production environment.

- **Lab environment**: Some organizations want the option to build an entire environment based on specific feature branches within their source code repository or to investigate specific gaps via a multi-day Agile/Scrum type technical or functional spike, or for specific innovation initiatives. These environments can be referred to as lab environments.

- **Beta environment**: Some organizations might choose to invite some of their customers or employees to use a beta environment to get high-quality feedback on new functions, features, or services.

- **Production**: A production environment is a live environment where the software can create business value by being accessed by end users. Here, the software is expected to be stable, reliable, and performant. The production environment is typically highly controlled and monitored, with strict security and compliance policies in place to protect the software and its users. That said, we expect the aforementioned production-like environments to be connected to the same monitoring systems and the same compliance policy applied and enforced.

Depending on the workloads, all these environments and their associated operations have to be duplicated for on-premises, co-location, public cloud, private cloud, and edge locations. The permutations across the environment's type and location can lead to a high number of different environments.

This requires automation and self-service automation provided by the platform team to be given to the tenants so that they can establish a required consistent enterprise-grade security and compliance posture.

Networking

The networking dimension needs to be informed by your tenant workloads, as well as the distributed technology environments. For example, do your targeted workloads run in containers and pods, and do those pods need multiple networking interfaces? Or do those pods depend on a service connectivity layer or a specific implementation thereof, such as a service mesh? If many applications depend on a service mesh, is that something you want to provide as part of your distributed cloud platform for your tenants so that they can reduce their cognitive load? Or is the complexity or compromise too big across the target workloads so that it makes little sense? Do you have confidential compute workloads you need support that require network segregation?

Where is the data located? Is data retrieval time-sensitive? Is our distributed data access synchronous or asynchronous or done via a service mesh? How time-sensitive are the updates to your data entities and how does the network contribute to failure scenarios and latency?

Do you have requirements for high-performance networking or network function virtualization through SR-IOV? Do you need overlay networks or different network segments for your control plane? Or do your workloads depend on **data plan development kit** (**DPDK**) libraries because they offload packet processing from the kernel space into the user space?

Are you executing cloud migration projects and have different application components running in different locations and require something such as Skupper, Submariner, Nexodus or Red Hat Application Interconnect?

Do you use a **content distribution network** (**CDN**) with static or dynamic data?

Does your distributed cloud estate rely on global traffic managers? Where do your load balancers sit and do they need to be redundant?

Are parts of your distributed environment managed by an operational technology group that has a network with a separate and heterogeneous technology stack? Are the interfaces proprietary or open source and open standards-based?

While there are often defaults for networking setups that you can choose from for particular services, such as gateways, subnets, security groups, and policies, they probably can't be applied across on-premises, edge, and cloud deployments. So, what is a fire-and-forget topic in a single cloud can evolve into a pretty complex undertaking in a distributed cloud context.

For telecom-related topics, network function virtualization is a key topic at the moment. Proprietary solutions from vendors with a long history and telco experience offer containerized and virtual network functions as part of their **virtual radio access network** (**vRAN**) and cloudRAN (that is, a mobile wireless network) product line. The open source alternatives concerning standards and product offerings are governed by the O-RAN (O for Open) initiative, with a large amount of established vendor support.

Other networking dimension-related concerns are east-west and north-south traffic, egress and ingress rules, network design and required subnet size, network zoning, network policies, NTP/time synchronization, proxy deployment, load balancing, best-fit **software-defined networking** (**SDN**) selection, and encryption termination points.

Storage

Storage is another dimension that can yield a great return on investment. In a co-dependency relationship with data, storage-related security and cross-location replication features, as well as non-functional requirements such as availability, throughput, latency, and capacity, demand a proper investigation.

While hyperscaler proprietary storage services look like an easy and good option in a single cloud context, a distributed cloud context needs further examination. Software-defined storage options and cross-site storage replication features for file/block and object storage from public cloud to co-lo (colocation)/on-premises to the edge can help keep cross-location storage operations, processes, and skills consistent while also providing the flexibility to change the master storage location. For example, a large car manufacturer found the public cloud storage bills to be unsustainable and hence was applying data filtering at their edge locations, as well as examining alternative CapEx-based master storage locations at their co-lo, while periodically erasing the public cloud storage repositories after replicating the storage off-public cloud.

Additional aspects of the storage dimension are storage life cycle management, including backup and recovery, and storage classes (including their associated performance and cost).

Platform operations, disaster recovery, and business continuity planning

RPO and RTO are terms that are commonly used in the context of disaster recovery planning and business continuity management.

Customer SLAs and potentially subsequently defined SLOs and SLIs are good for **business-as-usual** (**BAU**) operational scenarios. However, when disaster strikes, the focus shifts to RPO and RTO.

RPO is the maximum amount of data that an organization can afford to lose in the event of a disaster or outage. This is usually expressed as a time value, such as "we can afford to lose up to 5 hours worth of data."

RTO is the maximum amount of time that an organization can afford to be without a particular service or system in the event of a disaster or outage. This is also expressed as a time value, such as "we need to be back up and running within 2 hours."

Both RPO and RTO are important considerations when designing backup and disaster recovery strategies and the associated operations to ensure business continuity. It determines how frequent backups need to be taken to ensure that the organization does not lose more data than it can afford to, as well as how quickly the organization needs to be able to restore services to avoid significant disruptions to business operations.

Both RPO and RTO are used to help organizations plan and prepare for disasters or other disruptive events that could impact their operations. By understanding RPO and RTO, organizations can design backup and disaster recovery strategies that meet their specific needs and ensure that they can recover from disruptions as quickly and effectively as possible.

In a hybrid and multicloud context, this can be addressed through redundancy, active-active failover setups or fully-automated rebuilds, depending on the RTO and RPO. At the edge, we often don't have the luxury of redundant infrastructure and hence need to design different remediations.

With one of my previous employers, we planned to switch to a minimal order-to-cash process in the case of a multi-hour systems outage. That meant we had to plan for switching order and inventory taking, warehousing and transport logistics to paper-based processes, as well as plan how to switch back once the failing systems and integration points were back online. That alone required a massive amount of subject matter expertise and time to understand and document the process and backup procedures. Fortunately, we didn't have to go through that experience, but I am quite sure that we missed parts and probably underestimated the manual reconciliation work necessary in terms of data entry and system resynchronization as part of reestablishing normal operations. COVID didn't exist back then, so we missed the fact that we also needed to plan for a scenario where the systems were all available but people couldn't meet face to face, as well as for a scenario where both systems and people were not available.

We talked about antifragile in *Chapter 1*; this is exactly what we are referring to: scenarios where thinking about robust and stable practices are flawed and just not good enough because we don't know what we don't know, and hence what is going to hit us next.

So, what does that mean for our operating model? From a technical architecture perspective, we need to consider on-demand workload relocation and cross-location data access, including the underpinning infrastructure and public cloud service's service levels and how we architect around insufficient SLAs for services we have no control over. Redundancies, automation, potential process change, and reconciliation, as well as everything in light of the impacted business domain processes such as the aforementioned order to cash, can be expanded to all process domains, including but not limited to, order-to-activation, mine-to-ship, call-to-service, accounting-to-reporting, forecast-to-plan, hire-to-retire, and many more as part of non-functional requirements.

Documentation

"The code is the documentation" is a term you hear from people who find writing documentation boring – and many probably do. That might work for a team of Node.js developers, but not in a distributed cloud context.

Properly written, tested, and catered for, and your tenants as the focus target audience, the documentation provided by the platform and tenant experience stream needs to help tenants to reduce their cognitive load and make onboarding consistent and easy. Documentation with properly written and up-to-date instructions is important. It helps ensure the right behaviors are introduced, taught, and shared.

Establishing the wanted "right" behaviors is also beneficial from an adoption-at-scale and feedback loops perspective. You want your change agents to show the right behavior to all the people they help (that is, follow instructions) but also be able to point noobs to a set of up-to-date and up-to-scratch documentation to enforce and reinforce the right behavior. This enables adoption at scale but also learning to occur for the team that owns the platform and tenant experience through the feedback loops it helped establish. Feedback is valuable if it's in the context of the instructions that have been provided because you know what steps were followed by your tenant. This gives you a starting point to improve upon. As an example, do you want your tenants to follow planned processes and entry points when they try to get their development environment set up, use the provided self-service capability, and raise **feature enhancement requests (FERs)** for missing functionality or, on the other hand, start with skunk-works and create another user account and expense claim credit card bills?

A point to be aware of is that your target audience for edge workloads might use different language or business contexts or policies compared to your hybrid cloud users. If that's the case, ensure the documentation is understandable and can be followed from within different contexts.

In short, documentation and the associated training effect can be much more than just offering minimal instructions. It can showcase that you understand the needs of your tenants. It also allows you to gauge what works well and what doesn't work for your tenants. In turn, this will determine if your target operating model adoption is successful and going to plan or not. It is also a good practice to let your tenants share their feedback on documentation with comments/rating options directly in the document, which helps you understand the usefulness of the content and address the gaps diligently.

When I was working with Kevin Behr, the co-author of *The Phoenix Project*, one of his sayings was that the job at hand must "*make the right thing the easy thing.*" I think he hit the nail on the head.

To wrap this dimension up, we would like to take the *the code is the documentation* phrase further. What started as a tongue-in-cheek comment is taken up by organizations that document their wanted behavior as code – if you want to know what certain operational aspects entail, how backups are performed, or what compliance policies are applied to in your development environment, then you can study the automation playbooks in your company's code repository to find out. This is similar to how the OpenAPI specification led the way to help you generate code from your API contracts.

Furthermore, some open source vendors started to make what they call **validated patterns** available, meaning certain multi-product and distributed hybrid multicloud deployment best practices are available for everyone interested in reusing them via a simple Git pull.

An associated concept is operate first, which attempts to help organizations who want to learn how to properly navigate the operational concerns of a hybrid multicloud environment in alignment with their target operating model but with the help of leading practices.

Funding and budgeting

This dimension is about defining whether you want your platform as a product versus as a project, plus maintenance (which is most certainly the end of your cloud and edge efforts) and who is going to pay for it. Do you need to prove business value via show-back or charge-back? Or do you want to move into the world of FinOps and establish some core fundamentals around it? Let's look at some of these fundamentals:

- Cross-team collaboration
- Everybody accepts cost as their metric
- A centralized CoE drives DevOps
- Fast feedback loops, visibility, and reporting
- Business value drives hybrid multicloud and edge usage

I don't know about you but that whole FinOps movement still looks a bit half-baked to me and perhaps is a well-named distraction maneuver to draw attention away from the real problem, which is that utilizing public cloud providers is not necessarily reducing cost. Anyway, we'll leave it up to you to decide whether CapEx is actually not as bad as the cloud driven hype toward OpEx suggests.

If you believe in FinOps as the way to go, you should consider engaging with some consulting powerhouses such as Deloitte or Accenture, who have developed a FinOps capability map, to make FinOps part of your operating model. You may also consider leveraging finops.org as a starting point for your FinOps journey.

FinOps or not, more importantly, you need to determine who pays for your distributed technology platform, whether it's the tenants – who might then moan and complain about why it's so expensive and tell you that it'll be cheaper if they do it on their own (which of course is not true and mostly stems from the fact that small teams don't have full visibility of the compliance and security requirements at the enterprise level) – or a central IT budget, which most likely requires adequate reporting to justify increasing costs (OpEx) over the years, as well as developing the necessary tooling and compliance controls around it (CapEx) or a shared model where the CapEx is funded by the organization under the central IT budget and the OpEx is covered by tenant chargeback.

External dependencies

One of the five-time thieves in Dominica DeGrandis' book *Marking Work Visible* is called *Unknown Dependencies*. And what is true for software development projects is also true in a more complex distributed technology context.

Dependencies that inform the operating model are plentiful, so it's worth creating awareness of all first (self, internal), second (partner), and third-party (external, non-party entities) dependencies.

What you classify as first, second, and third-party dependencies varies based on your industry and partner network. So, please just take this as a guide to tickle your brain into thinking.

Examples of first-party dependencies are in-house developed tools and automation scripts, libraries, enabling platform and facilitating teams, budgets, in-house training, and talent acquisition. Another example is a team you need to synchronize your release train with.

Examples of second-party dependencies are all your partners, who could be providing tooling, developers, consultants, components of your products, or services such as database backups.

Examples of third-party dependencies could be COTS applications, IoT devices such as sensors, microcontrollers manufacturers such as Arduino, ARM hardware suppliers, GPU manufacturers, edge devices, edge location providers, AI foundation model providers or camera manufacturers that supply you with devices with built-in AI capabilities, machine learning and other specialized sub-system team resources, and energy and data center providers.

Other things to consider are DNS, CDN and global traffic managers, IAM, monitoring, logging and alerting tools, internet access, external asset registries, CMDBs, and IPAM.

Summary

In the *Platform and tenant experience stream* section, we provided some suggestions regarding dimensions and the questions you should ask yourself to define your guiding principles (soft guidance) and the guardrails (hard rules) with your tenants while having an overall "platform" experience in mind. Perhaps those questions might open that can of worms we all know about. This is not a bad thing – it just means you discovered a part of your business that needs more attention and you won't have to clean up an unintentional mess afterwards. Questions might reveal an area of your business that's so much deeper and so full of specialized subject matter that it qualifies for another dimension in your operating model.

In the next section, we will dive deeper and focus solely on the workloads, applications, products, and consumable services that your tenants run, which can be a mixture of revenue-generating or vital to your business functionality and a simple but non-vital productivity app for your field service teams.

Workloads and applications

The third and final stream is the workloads and application stream, which some organizations might more appropriately call products and services if that's the language that's used within the organization.

In this category, you can further distinguish between inner and outer loop streams if you want to introduce sub-streams. Going forward in this book, we will keep the loops as our two core dimensions. The loop dimension's purpose is to do the following:

- Reduce cognitive load
- Optimize productivity
- Establish the foundation for evolving compliant and enterprise-ready application delivery and deployment:

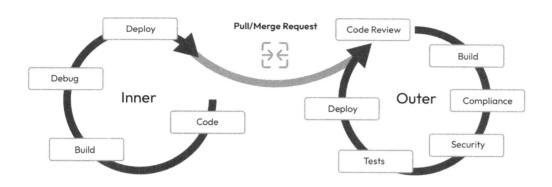

Figure 5.11 – Inner and outer loop concerns

The outer loop

The outer development loop is used for production releases, such as code merges, code reviews, testing, deployment, and the release process itself, including canary releases and A/B testing – which is more a market research approach than a release at its core, though it still affects release processes, dark launches, and so on. The outer development loop also includes automated CI/CD pipelines and GitOps workflows:

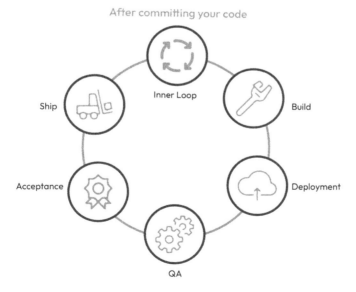

Figure 5.12 – An outer loop example

The outer loop dimensions require a close platform and tenant experience stream stakeholder engagement. It is the glue between the inner loop and the platform team and is where the rubber hits the road in terms of the resulting developer experience. A good outer loop stream lead is someone with good relationship-building skills and background on both sides of that structure, which consists of platform engineering and product teams.

The main requirements of the outer loop are to ensure that the latest code and configuration changes are as follows:

- Always buildable (continuous integration)
- Always deployable (continuous delivery)
- Automatically deployed to the target environments (continuous deployment – not always implemented)

There is also a space between the inner and outer loop beyond technology. Said differently, some topics cross the inner and outer loops the same way as there are topics that cross the different streams. There is also a healthy friction between what the platform provides organization-wide, what the outer loop provides team-wide, and a further distinction between what the developers need and what they want.

Examples of such boundary crossing topics are shift left and flow metrics.

Shift left – if misunderstood – can be fraught with danger. Just piling more and more things to worry about ("shit", short for shift left, is the technical term here) onto a developer's plate is not what "shift left" intends to do. Ideally, your operating model should create an experience that makes sustainability, security, and compliance concerns transparent to the developer while still addressing them early in the **software/system development life cycle (SDLC)**.

Let's look at some examples:

- **Shift left done wrong**: Send an email to all developers stating that they can no longer download container images from Docker Hub due to compliance and security concerns.
- **Shift left done right**: Establish an organization-wide container registry and make only supported, compliant, and certified container images available for your developers to use. Point them to accessible documentation and "Getting Started" guides. Explain the why, what, and how.

If you can do this, then you will solve your security problem without additional cognitive load on developers and with minimal impact on productivity.

The inner loop

The inner (development) loop is what the individual developer does. A developer lives in the inner development loop to code and test changes quickly. From an intrinsic motivator point of view, this is where organizations need to focus on making individuals feel productive, that they are contributing, and that they are making an impact.

The inner loop is about defining the local developer experience, made up of the following:

- **Integrated development environments** (**IDEs**) configured to the organization's standards

- Having a runtime environment, as supported in production

- Debugging tools best fit for purpose found through team consensus

- Container images as per corporate standards that are supported and certified with SBOM and provenance

- Libraries that are compliant with software supply chain requirements

- Programming languages, as supported in production

- Test frameworks

- Peer code review

- Specific workload requirements, such as confidential computing, high-performance computing, and quantum computing

- Environment simulators for quantum, Radio Access Networks (Telco), the edge, and operational technologies such as IoT sensors, ARM processors, RISC-V, FPGAs, and more

- SDLC stage-relevant data access with relevant data security and compliance controls in place

- The developer's flow metrics and the time thieves we discussed earlier for examining unknown dependencies, including too much WIP, unknown dependencies, unplanned work, conflicting priorities, and neglected work

- xOps – whatever x is in your organization, be it DevOps, FinOps, GitOps, or other cultural approaches that ultimately impact the work of the developer

- Design guidelines for cloud-native, cloud-ready, and cloud-optimized workloads in general but also particular topics, such as for implementing the following:

  - IAM integration

  - Authorization

  - ZTA

- Suitable SLIs to meet SLOs

Application modernization and legacy workload migration

An important aspect of the cloud and edge journey is selecting the right workloads.

New applications (greenfield) are usually easier to develop than migrating existing legacy workloads (brownfield). This is mostly because existing applications support already-running and often important business processes that aid with risk reduction, cost optimization, and/or revenue generation. Those applications come with integrations to other applications and systems, as well as support staff who know where to look to fix business-critical issues as fast as possible.

We don't rate cloud migration slogans such as "*x apps in x days.*" This might be great for the hyperscaler marketing machinery or industry technology executives to generate a nice headline in the press, but unfortunately, we have seen those approaches fail more often than they succeed.

A reasonable approach often starts with an application portfolio analysis. The more applications you have, the more you wish to prioritize. Piloting helps you learn vital lessons quickly by testing solutions to challenges you encounter. You can start by piloting the migration or modernization of strategically important applications of medium complexity with the relevant scalability, resilience, and high availability requirements; basically, you choose workloads that benefit from an elastic cloud environment. Avoid low-hanging fruits like the pest. If you can't solve the challenges of medium-hard or medium-complex workload migrations/modernizations, then starting with low-hanging fruit just creates a mess in the long term or unnecessary costs if you only find out later that there are non-circumventable show-stoppers.

Ideally, even your legacy applications should adhere to the low coupling and high cohesion principle. This could mean that Java applications, for example, can be broken up along existing interfaces.

While there might be an eagerness to port all applications to the edge or a public cloud, it's also wise to step back and check whether it is worth the effort for applications that are a much better fit for non-scale out environments or are not suitable for low-powered edge environments.

The following is a classification that we recommend:

- **Non-cloud/non-edge**: Applications that don't benefit from running on a cloud or edge environment, have a low ROI, or require heavy lifting to be ported to run in a cloud or edge environment
- **Cloud-ready/edge-ready**: Applications that can be migrated to a cloud or edge environment with minimal modifications
- **Cloud-optimized/edge-optimized**: Applications that were modified or enhanced to efficiently run in a cloud or edge environment by leveraging cloud-specific features to improve performance, scalability, and cost efficiency
- **Cloud-native/edge-native**: Applications that have been developed, modified, or enhanced to efficiently run in a cloud or edge environment

We believe that an objective common-sense approach will yield the best results.

Summary

In this chapter, we provided a comprehensive approach in terms of defining your own distributed technology operating model. We listed many concerns and posed even more questions for you to consider, which you hopefully find suitable for your organization.

You are now equipped not only with open principles and practices that help create a high-performance culture but also with a structure that you can apply to define your organization's best-fit operating model.

In the coming chapters, we will show you how to create your operating model based on an example, while also covering some of the challenges you can encounter and how you can address them.

As a final note, whatever implementations, tooling, or engineering you do within any of your operating model dimensions, I invite you to consider contributing back to an active open source community. That not only establishes you and your organization as a cloud and edge leader, but it also helps raise the tide and hence all other boats. As a reward, you may get an open source vendor to take your development into their product roadmap, which means it takes the support burden off you and reduces your technical debt.

Further reading

Refer to the following links for more information about the topics that were covered in this chapter:

- McKinsey on operating model iteration: `https://www.mckinsey.com/capabilities/mckinsey-digital/our-insights/how-to-start-building-your-next-generation-operating-model`

- Ron Westrum – culture topologies: `https://www.ncbi.nlm.nih.gov/pmc/articles/PMC1765804/pdf/v013p0ii22.pdf` and `https://www.ncbi.nlm.nih.gov/pmc/articles/PMC1765804/`

- Cultural practices – Ron Westrum's research applied: `https://itrevolution.com/articles/westrums-organizational-model-in-tech-orgs/`

- Dan Young and hybrid cloud shared responsibility models: `https://www.redhat.com/architect/shared-responsibility-model-srm-hybrid-cloud`

- *Making Work Visible*, by Dominica De Grandis: `https://itrevolution.com/product/making-work-visible/` and podcast `https://podcast.andreasspanner.com`

- Team topologies: `https://teamtopologies.com/`

- The four golden signals of observability: `https://www.ibm.com/garage/method/practices/manage/golden-signals`

- BCG on the distributed and federated future of data architectures: `https://www.bcg.com/publications/2023/new-data-architectures-can-help-manage-data-costs-and-complexity`

- Operate first: `https://www.operate-first.cloud/`

- Validated pattern: `https://hybrid-cloud-patterns.io/patterns/`

- Data architectures: `https://www.bcg.com/publications/2023/new-data-architectures-can-help-manage-data-costs-and-complexity`

- The PRISM program: `https://en.wikipedia.org/wiki/PRISM`

- *Postmortem Culture*, by Google: `https://sre.google/sre-book/postmortem-culture/`

6
Your Distributed Technology Operating Model in Action

In the previous chapter, we learned how to create our best-fit distributed cloud and edge operating model. We followed six main steps to determine our operating model dimensions for each operating model stream and, subsequently, our complete technology operating model. We placed our stakeholders into stream teams and grouped dimensions under each stream, which the stream teams own, and subsequently built out an operating model. We discussed how to create stakeholder mappings and how to use them to run workshops to build the operating model. We then elaborated on each dimension. In this chapter, we will put the operating model concept to use. By the end of this chapter, we will have learned how to do the following:

- Build a distributed cloud and edge technology operating model while following the process defined in *Chapter 5*, for an insurance organization based on a simulated case study

- Use the Open Practice Library to build the operating model

The following topics will be covered in this chapter:

- Introducing a fictitious case study of an insurance company, including competitive threat, global expansion plans, and a digital transformation initiative

- Building a distributed cloud and edge operating model for the aforementioned organization

- Implementing and operating your business applications and technology estate based on the technology operating model you created previously

We'll start this chapter by looking at the case study that we will be using in this chapter.

Simulated case study

A fictitious automobile insurance company expands globally, acquires a home insurance company, adopts modern cloud and edge application development and delivery practices, and attempts to transform into a digital business.

Case study overview

Bison Insurance is a well-established automobile insurance company in North America with a long history of providing high-quality products and services to its customers. Bison Insurance started as an automobile insurance company in 1934 in Chicago, covering Illinois and Milwaukee. By 1948, they were firmly established as the number one automobile insurance company in these two states. They had over 300 physical stores and built a strong freelancer network to take their services to the rural markets. By 1950, they enjoyed remarkable customer loyalty with a 97.8% renewal rate while continuing to acquire new customers at a rapid rate. As the economy grew from strength to strength post-war, Bison Insurance grew faster and expanded to other states, both organically and via acquisitions. As of 1965, they are operational in all of the United States and are the number three automobile insurance player in the country. Bison Insurance went public in 1968 and had tremendous success in the first 3 years of going public, with this share price growing at an average of 27% year over year.

The 1970s threw the first big challenge at Bison Insurance. Rising fuel prices, combined with the recession, saw auto ownership fall by as much as 20%, delivering a significant impact on their revenue and stock value. During this difficult time, Bison Insurance halted expansion and focused on creating operational efficiencies. They needed to become leaner to deal with the tough economic conditions. They also postponed a planned diversification strategy until the economy rebounded. This strategy involved introducing new lines of business to access new markets.

The 1980s brought better fortune for the company. As the economy came out of recession and middle-class affordability grew, Bison Insurance's revenue and stock value increased. They still relied primarily on physical stores and freelancer channels to drive their growth. With an overall healthy market for automobile insurance, Bison Insurance grew faster and recovered its market position, and also saved enough cash reserves to embark on the next phase of its journey.

As part of its diversification and growth strategy out of the 1970s, Bison Insurance has been exploring expansion opportunities in global markets and adjacent lines of business. They acquired a home insurance company in North America called Holstein in 1989. Holstein is one of the top home insurance companies in North America, with a top three market share in the USA and Canada. They are a private company with healthy profit margins and growth. The financial details of this transaction are kept private.

In 1993, Bison Insurance also acquired a European insurance provider called Wisent. Wisent is a popular insurance provider with both automobile and home insurance services in Western Europe. Wisent provides agricultural insurance in the UK, Germany, and the Netherlands. The acquisition was completed in 1994, but Wisent is still operated as an independent organization.

In 2002, Bison Insurance rolled out new in-car device for tracking driving behavior called Gaze in the USA and later expanded to Canada in the following year. Gaze helped them provide better customer experiences and personalized quotes to attract new customers. However, operationally, Gaze has been an expensive product to maintain, with the device needing to be serviced or replaced every 18 months and sometimes being damaged during small accidents. Given this, Bison Insurance is cautious in providing Gaze devices, only providing them to existing customers upon renewal.

Around this time, Bison Insurance also ventured further into technology as e-commerce became mainstream and a growing number of their competitors started providing customer portals as a self-service channel for their customers. For the North American market, they launched a web portal for automobile insurance and another one for home insurance. Later, based on customer requests and marketing research on potential cross-selling opportunities between automobile and home insurance customers, Bison Insurance combined these portals. During this time, the functionality of the portal also improved from just reporting accidents to a full-service portal with bill pay, contract signing, and more. The success of the BisonInsurance.com portal helped the organization reduce its dependence on physical stores and streamline its freelancer agents' network interactions with dedicated agent modules in the portal. Bison Insurance is on track to close 98% of physical stores by the end of 2025.

In Europe, Wisent followed a similar journey with varying degrees of success. For automobile and home insurance, the adoption has been high and they were able to seamlessly migrate their customers to the digital platform. However, agriculture insurance customers remain resistant to using the digital platform. These customers still prefer in-person/telephone communication over the self-service portal. The following figure shows the high-level timeline of Bison Insurance's journey, including its expansion to Europe with the acquisition of Wisent:

Figure 6.1 – Timeline showing Bison's expansion across USA and Europe

As shown in the preceding picture, from its humble beginnings, Bison Insurance has grown to be a global business with organic growth, portfolio expansion, and acquisitions.

The IT landscape

Bison and Wisent maintain their own IT departments in their respective regions.

Bison Insurance relied on **commercial off-the-shelf** (**COTS**) applications and a smaller IT team to maintain their back office and physical store systems. They also had an **interactive voice response** (**IVR**) system built by a local vendor in Chicago who continued to maintain the system for a few years.

During the dotcom boom, Bison Insurance expanded their IT team to add additional capabilities in development and project management to explore a new web portal initiative. As part of this, they also acquired a local boutique IT consulting firm called *Green Vision* in 1999, which specialized in building web portals. Green Vision brought in specialized IT skills that helped launch Bisoninsurance.com and also helped them build a new back-office insurance application for their physical stores. The IVR system was upgraded in 2006 as part of the telephony system upgrade. The internal IT team played a significant role in selecting the new IVR platform and worked with a vendor called Voicecom to build the necessary extensions and customizations to make it fit Bison Insurance. With this, they were able to maintain the new IVR system in-house with as-needed support from Voicecom. The Gaze device and the data transfer platform is a white-label service provided by a US entity called EdgeInnovations. EdgeInnovations takes care of all the logistics and device management on a per-device and per-ticket cost basis. They also have a process to sync up the data from the devices to Bison Insurance, which is backed via secure FTP every week. This data is ingested into the core system database via a batch process that runs daily.

Wisent has a relatively small footprint, with most of its IT functions outsourced to a regional IT firm called Magpie. Wisent has a small internal IT team that takes care of primarily project management, architecture, and IT operations functions.

From an infrastructure perspective, Bison Insurance and Wisent manage their own regional data centers. They have a very small cloud footprint for data backup and workplace solutions. Bison Insurance owns two data centers in North America, one in their Chicago office and another in their Toronto office. EdgeInnovations uses a cloud provider to host their Gaze platform, but since the Gaze platform is completely managed by EdgeInnovations, Bison IT is not directly involved in the day-to-day operations of this cloud environment.

They follow a similar model for IT teams as well. Both Bison Insurance and Wisent own their own independent IT teams. Bison Insurance has a larger IT team with in-house project management, architecture, development, and delivery capabilities. On the other hand, Wisent has a smaller IT team with project management and architecture capabilities. They rely on Magpie for most of their IT needs. Both Bison Insurance and Wisent follow the traditional waterfall development methodology and the ITIL service management framework to deliver their IT services. The testing, security, and compliance processes are also mostly manual and sequential.

Overall, Bison Insurance and Wisent can support their business requirements with their current IT setup through a combination of in-house IT teams and external partners. They are capable of quarterly minor releases, with new features released every 6 months for their homegrown applications. This aligns with their current business expectations. The following figure shows the IT landscape across Bison Insurance and Wisent:

IT landscape

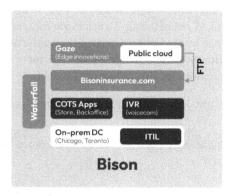
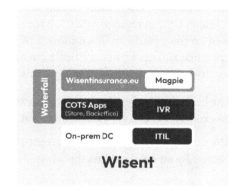

Figure 6.2 – Current Bison and Wisent IT landscape

As shown in the preceding figure, both function with little commonality across technology, partners, and processes.

Challenges and threats

Bison Insurance is facing significant challenges as it moves into new markets and adopts new technologies. New data sovereignty requirements are rolled out in various regions across the globe, which creates additional complexities to their expansion plans.

Previously, both Bison Insurance and Wisent only used on-premise data centers to support their North American and European business. However, with their global expansion and new acquisitions, they are looking to move to the public cloud to leverage the benefits the cloud ecosystem offers and to reduce internal IT dependencies. This will require a significant investment of time and resources, and will also require a complete overhaul of their technology stack and modernizing their existing applications to leverage the best capabilities the cloud environments have to offer. Additionally, Bison Insurance needs to revisit its development and operation models, moving from waterfall to agile methodologies to keep up with the pace of digital startups that are venturing into this space.

One of the biggest threats that Bison Insurance faces is from a digital startup called Sharps. Sharps is a fast-growing and born-in-the-cloud digital organization that is disrupting the insurance industry with its innovative technology and customer-centric approach. Sharps is growing at twice the rate of Bison Insurance, with cost-effective options and better customer experiences. This has made them very popular among younger, tech-savvy drivers who are looking for an alternative to traditional insurance companies. Their sustainability messaging is also resonating well with millennials and Gen-Z customers, who believe that by signing up with Sharps they are doing good for the planet by reducing their carbon footprint. Sharps is piloting spot insurance options where people can take short-term auto insurance from them for car rentals.

Another digital startup that is gaining popularity in this space is SelectTheBestInsurance.com. This startup is building a marketplace for customers to compare insurance quotes from different insurance companies for auto and home insurance in North America. They are growing steadily and are marching toward 1 billion USD in revenue by 2027. At present, they have 7 out of the top 10 insurance companies onboarded on their platform. Even though Bison Insurance wanted to be listed on the platform as soon as possible and made this a critical item on the IT priority list, it is struggling to implement the API integration that is required to achieve this. The main reasons for this failure are twofold:

- The backend core business application architecture and data model do not lend themselves to be exposed as APIs for external integration
- The network architecture and current security standards do not allow external access to the backend systems/data directly

This issue had made it to the top and the CEO Ed Coolidge was asked to share his findings and how to address this in the last board meeting.

Proposed next steps

Ed worked with the CIO Ian Hornaday and the CFO Fiona Dawes to come up with a strategy to address this issue and also new competitive threats that emerge from these high-growth, fast-evolving digital startups. Ian tasked his enterprise architecture team to work on this and come up with a report that he presented to Fiona and Ed. Ed and Fiona were impressed with the report and quickly worked on a financial model and presented it to the board. The following are some of the key approaches that were highlighted in the board presentation:

- Create a centralized IT team for both Bison and Wisent and make Ian the global CIO.
- Make the current CIO of Wisent, Donna Olech, the Chief Digital Officer, who focuses on digital transformation. Donna will report to Ian but also have a dotted line to Fiona and Ed to provide direct updates every month.
- Create a new program called "Resurgence" and allocate a 175 million USD budget for this program for the next 5 years initially.

Under the **Resurgence** program, achieve the following outcomes:

- Move from a legacy on-premise data center and operating model to a hybrid cloud and edge architecture and cloud operating model to reduce internal IT dependencies and maximize flexibility.
- Upskill the internal IT team to accelerate application development and delivery by following modern cloud-native application development, delivery, and automation practices.
- Reduce dependency on external partners and build new IT capabilities in-house to be able to build, deliver, and maintain applications.

- Create a modernization strategy for the existing business application and modernize high-value business application and retire low-value apps. Consider leveraging SaaS platforms as much as possible as part of this exercise to ensure only key business differentiating applications are built and maintained in-house and that other apps are purchased from ISVs, preferably as SaaS.

- Revisit the partnership with EdgeInnovations for the Gaze device program, which is a costly and less popular option. Smart cars can deliver these metrics out of the box and may be willing to share this information with insurance providers in the future.

The board reviewed the presentation and gave a thumbs up in the following week for the Resurgence program. Donna was announced as the executive program sponsor and she immediately started working on this. The following figure shows the new org structure that was created for the Resurgence program:

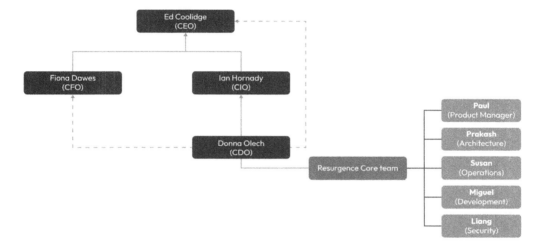

Figure 6.3 – Resurgence program org structure

In the following section, we will show how Donna built a team and followed the operating model blueprint that we introduced in *Chapter 5*, *Building Your Distributed Technology Operating Model*, to create and implement a distributed edge and cloud operating model under the Resurgence program.

Building an operating model for the Resurgence program

In this section, we will see how Donna and the team built their new operating model following the six-step process highlighted in *Chapter 5*.

First and foremost, there are a few important foundational things that Donna did to build a core team that will be driving this initiative. Donna chose members from different parts of IT teams across the enterprise architecture, development, operations, and security teams. A total of four members from these teams, one from each (Prakash from enterprise architecture, Miguel from development, Susan from operations, and Liang from security) were identified for the Resurgence program and asked to work full-time on this initiative (please refer to the previous figure for org structure). Donna also hired a new program manager called Paul Edenberg for this initiative. Paul comes with more than 20 years of program management experience in the insurance industry across the UK, US, Middle East, Australia, and Singapore, and in his recent stint is leading a major transformation project for a global insurance player in the USA. Donna also allocated an internal communications budget for this initiative to create organization-wide awareness and focus. With this in place, she requested Ed to make an organization-wide announcement about the program. Ed announced the Resurgence program, Donna's new position and her team's involvement, and the importance of it for the future of the company.

With the announcement and the Resurgence core team in place, Donna requested the core team to initiate the program. The team agreed on the following six-step approach:

1. Identify communication and collaboration tools while meeting the cadence and documentation requirements, as well as a repository.

2. Document the current IT landscape.

3. Create an internal communications plan.

4. Build the distributed cloud and edge operating model.

5. Implement and operate the distributed cloud and edge operating model.

6. Ongoing cadence and improvements.

While Paul took ownership of tasks 1, 3, and 6, the core team has been tasked with tasks 2, 4, and 5. For this book, we will focus on tasks 2, 4, and 5, which are owned by the core team.

Documenting the current IT landscape

The first critical task the focus team worked on was capturing the current IT landscape across Bison Insurance and Wisent; most of the information was already captured across different documents and asset registers. The goal of this exercise is to consolidate all IT assets and their corresponding business value across the following areas:

- **Infrastructure**: Servers, network, and storage

- **Applications**: Home-grown, COTS, and SaaS

- **Backend systems**: Databases, messaging systems, and so on

Given the representation of different IT teams in the mix, the core team members started collecting existing documentation from their respective teams. They also leveraged existing collateral that was available with the PMO and EA teams as well. They were able to get most of the information they were looking for. However, the information was siloed in nature across Bison Insurance and Wisent, and given the way the teams were structured and operated, every team had documentation on the services they offered, along with the infrastructure/applications they maintain. The core team had to work with the PMO team, EA team, and LOB to map all the technology assets to the business functions to the business value they deliver.

Once the IT assets were captured, the core team had a good understanding of the current IT landscape, which serves as the foundation for the next step in building the operating model.

Building the distributed cloud and edge operating model

As we explained earlier, we will be using the six-step process defined in *Chapter 5*, *Building Your Distributed Technology Operating Model*. The following steps were identified:

1. Define the streams.
2. Manage stakeholders.
3. Select dimensions.
4. Scope the dimensions.
5. Detail the dimensions.
6. Review and iterate.

The following figure shows the process we use to build the operating model using OPL practices:

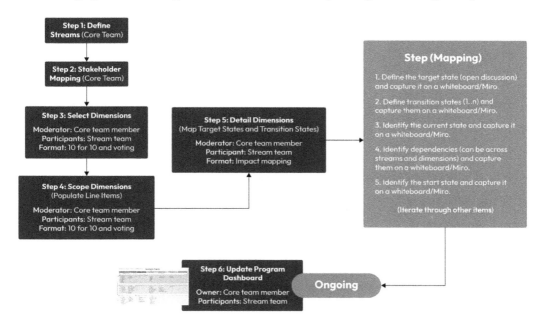

Figure 6.4 – Process used to build the Resurgence program's operating model

In the following sections, we will see how the core team defined the streams as the first step.

Defining streams

As explained in *Chapter 5*, *Building Your Distributed Technology Operating Model*, defining streams is important to improve the overall efficiency and manageability of this workstream. For the Resurgence program, the core team started by defining the streams based on the objectives of the program that were listed earlier in this chapter. Given that the Resurgence program impacts all major business and IT functions across Bison and Wisent, all three streams are important. The finalized streams are as follows:

- Leadership and architecture
- Platform and tenant experience
- Applications and workloads

With the streams identified, the core team moved on to selecting the stakeholders for each stream.

Stakeholder management

To ensure the right stakeholders were assigned to the streams and to ensure their ongoing participation, the core team leveraged the existing org chart, application owners, and IT team structure that were documented under the application portfolio matrix and the current and future projects pipeline by the PMO and enterprise architecture teams. They focused on creating a balance across Bison Insurance and Wisent between various team representations:

- IT team
- Security
- Enterprise architecture team
- PMO office
- Finance
- Line of business application and process owners
- Senior management
- App support teams

Overall, they identified 20 stakeholders from this exercise. They leveraged a 2x2 matrix across the impact and influence vectors to identify the 20 stakeholders. The following figure shows the 2x2 stakeholder matrix that was developed by the Resurgence program's core team:

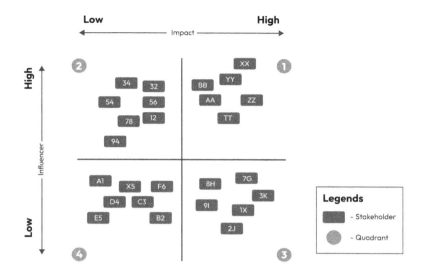

Figure 6.5 – 2x2 stakeholder mapping matrix

Once the matrix was developed, they focused on recruiting as many stakeholders as possible from quadrants 1, 2, and 3. Higher importance was given to stakeholders from quadrants 1 and 3, but necessary care was taken to include good representations from stakeholders in quadrant 2.

The core team conducted interviews with all the identified stakeholders in these three quadrants before recruiting them to make sure they were motivated and had enough bandwidth to participate in the Resurgence program. After, the stakeholders were grouped into three different streams based on their current role, expertise, and the interest they expressed during the interview. The following figure shows the grouping for the stakeholder streams:

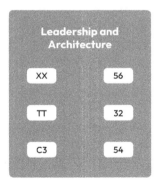  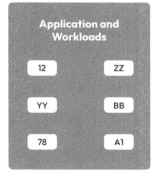

Figure 6.6 – Stakeholder streams mapping the stream team to the stream assignment
(it is possible to have overlap so that people can be part of multiple streams)

With the stakeholders identified and grouped into three streams, the Resurgence program core team moved on to the next stage, which was using more workshops to select the dimensions under the identified streams.

Selecting dimensions

With the stakeholder stream mapping done, the core team was ready to run workshops to work on the dimensions for the identified streams. They ran three sets of workshops; one set per identified stream, with three members from the core team to facilitate these workshops. The purpose of these workshops was to cover the following aspects of the streams and dimensions:

- Identify the dimensions
- Scope the dimensions
- Detail the dimensions

They started with a 3-day workshop that was run off-site to improve collaboration and reduce distractions, with a plan to extend it to remote workshops if needed. Some key techniques from the Open Practice Library that were highlighted in *Chapter 5, Building Your Distributed Technology Operating Model*, were utilized in this workshop.

In this section, we'll look at how Susan ran the leadership and architecture stream workshop. Susan started the workshop with a 15-minute icebreaker session to help participants learn about each other and to break initial barriers. *"Let's all do a warm-up exercise. Can you introduce yourself, your role, and what excites you to be part of the Resurgence program? If you could hang out with any cartoon character, who would you choose and why?"* asked Susan.

The team looked at each other and waited for someone else to go first. After a few minutes of silence, Michael started by saying *"My name is Michael Law. I am a senior director of IT at Wisent, based out of the UK. The thing that excites me the most about the Resurgence program is to be part of the future of the company, which has the potential to make us number one in the industry. I think I would like to hang out with Bugs Bunny. I have been a big fan of his sarcastic humor from a young age and would enjoy spending time with him,"* he closed with a smile. *"Thank you, Michael, I now know where your sense of humor comes from,"* said Susan. The other members followed by introducing themselves and also shared the cartoon character they would love to spend time with. The team loved the introduction from Juan. Juan said, *" I am Juan Gonzalez, vice president of the claims processing department based out of Chicago. I think the Resurgence program is going to revolutionize the way claims are filed and processed and I am looking forward to being a key member in shaping it. In terms of a cartoon character, I grew up in Chile and my favorite character is Condorito. I love the way he solves problems using his wit and I think I would learn a lot from him." "I have never heard of Condorito before, have any of you?"* asked Susan. It turns out that no one else had heard about Condorito either.

The second session that Susan ran is to establish a **social contract** as a team agreeing on five acceptable and five unacceptable behaviors during the workshop. The 10-for-10 brainstorming exercise from the Open Practice Library was used for this session. Every team member was given two sets of sticky notes in different colors – one set to write down the acceptable behaviors and another set to write down the unacceptable ones. The team was given 5 minutes to write down 10 acceptable and 10 unacceptable behaviors. They were then asked to take a minute to select the top five acceptable and unacceptable behaviors and asked to stick them on a whiteboard. With all six stream participants, a total of 30 acceptable and 30 unacceptable behaviors had been identified. In this case, some duplicates might have existed, so it would be a good idea to group similar ideas to make the next step easier.

As the next step, the teams were asked to identify the top five behaviors on each side via a 3-minute voting exercise. Each participant was given a maximum of 10 votes as round stickers (one vote per sticker). They could apply as many votes as they wanted based on how they felt about the particular behavior – that is, they could apply all 10 stickers to one behavior or two stickers per behavior, three in the acceptable and two in the unacceptable category, and so on. The facilitator then arranged the top five acceptable and unacceptable behaviors. This contract now acted as the guidelines for the workshop to follow. The following figure shows the social contract that was created for the leadership and architecture stream workshop:

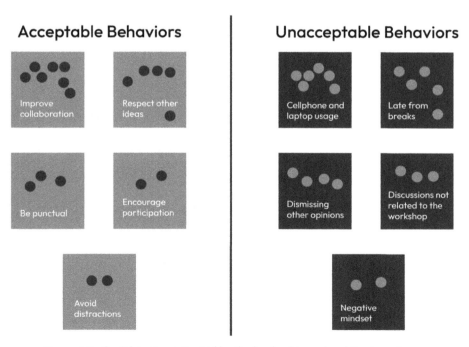

Figure 6.7 – Social contract created by the leadership and architecture stream

Susan went through the finalized social contract, printed a copy of the contract by taking a picture of the whiteboard, and asked all stream team members to sign it. She then said, *"Great going team – we are now ready to get a little deeper: we are going to start working on identifying the dimensions under the leadership and architecture stream. However, before we do that, let's take a 15-minute break. Please come back on time so that we can start the next exercise as planned."* The team left the room for a much-needed coffee break and was back on time. *"Welcome back, team. Let's start working on identifying the dimensions under our leadership and architecture stream. We will be following the 10-for-10 brainstorming and voting exercises that we used for establishing the social contract in the last session. As a reference, please use this example I have projected on the screen."* Susan pointed to the dimensions chart from *Chapter 5, Building Your Distributed Technology Operating Model.* The following figure shows the streams and dimensions dashboard that Susan used as an example:

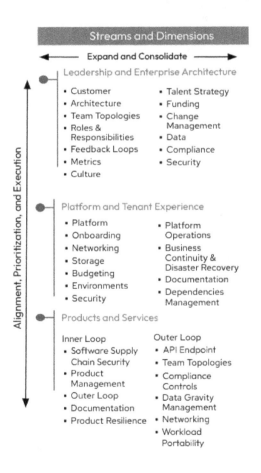

Figure 6.8 – Streams and dimensions mapping example

Susan then ran the 10-for-10 brainstorming exercise while following these steps:

1. Susan asked the participants to write down all the dimensions that needed to be included in their stream in 15 minutes on sticky notes.

2. She then asked the participants to stick their dimensions on a whiteboard (or sticky chart).

3. Susan then worked on grouping similar items, asking relevant questions to the participants as needed, and removing duplicates.

4. She gave the participants 10 circular stickers each (dots) and asked them to vote on the dimensions' starting order while keeping dependencies in mind. They were provided with a choice to allocate as many or as few votes as they wanted to their dimensions of interest. 5 minutes were allocated for this voting exercise.

5. Susan then rearranged the dimensions based on the votes they received on the whiteboard. She also facilitated discussions if a tiebreaker was required – for example, if seven dimensions are voted as two.

6. *Steps 4* and *5* were repeated until the stream team was satisfied that all the key dimensions had been identified and ranked appropriately.

7. Susan then populated the dimensions, along with their rank and the total number of votes, in the dimension sheet. All unvoted dimensions were either consolidated into existing dimensions or put into their own dimensions. At this stage, it was worth investigating why specific items didn't receive any votes as this might have been an indicator that an important aspect was not well understood or that the stream team was incomplete if the SMEs in the teams were missing. This process was followed by other core team members as well so that they could capture all unvoted dimensions and run a workshop with all three streams together to rank these items and move them to appropriate streams later as required.

> **Note**
>
> It is not recommended to restrict the number of dimensions. Based on your organization's need and the quality of the dimensions identified, the core team can either increase or decrease the number of dimensions selected under each stream. Fewer dimensions will most likely have a wider scope; more dimensions can end up with a smaller scope.

The following streams were selected for the leadership and architecture stream:

- Customer
- Partner
- Competitor
- Architecture
- Data

- Culture
- Security
- Compliance
- Talent strategy
- Metrics

A similar process was followed for the other workstreams; the selected dimensions are listed here.

Platform and tenant experience:

- Platform
- Onboarding
- Life cycle management
- Infrastructure standards
- Platform security
- Platform operations
- Business continuity and disaster recovery
- Documentation standards
- Dependency management
- Budgeting

Applications and workloads:

- Developer experience and tools
- Workload categorization
- Workload portability
- App modernization
- Data gravity management
- Applications and workloads documentation
- Technical debt management
- API economy

All the dimensions that were selected across all stream teams were then populated in the template, as shown here:

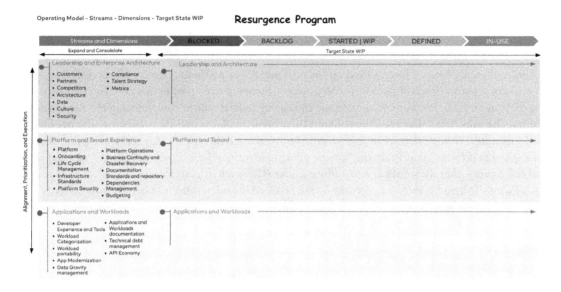

Figure 6.9 – Operating model template populated with the selected dimensions

This template acted as the foundation for the next few steps as the core team started scoping the dimensions and items within the dimensions across all stream teams.

Scoping dimensions

As the next step, the stream team worked on scoping the dimensions. It was a new day for the Resurgence program. The core team member, Prakash, needed to find the best way to get his team across the scoping dimensions exercise for the platform and tenant experience stream. The team was familiar with the 10-for-10 brainstorming and voting exercises, so Prakash was confident the team was ready. Prakash started the day with another ice-breaker exercise, a practice from OPL. The team started assembling and got seated with excitement to see how the day unfolded.

Prakash said, *"Welcome team, I hope you all enjoyed the last workshop where we selected the dimensions. Today, we are going to dive a little deeper into scoping the dimensions. We are going to follow the same model with 10-for-10 brainstorming and voting exercises for this. How about we start the day with an ice-breaker exercise? It is a simple process to warm us up for the day."* He continued, *"Let's share what our favorite place to visit is and why. Let me go first. My favorite place to visit is a particular tree on my ancestor's farm back in India. That coconut tree is supposed to be more than 75 years old. My grandfather planted it when he was 10 years old as the first tree on the farm. According to my grandfather, it is a lucky tree and helped his father, my great-grandfather, to establish a farm and raise 11 kids – no kidding! – and kick-started the revival of our family from poor peasants to the middle class with the ability to afford good education and living conditions for us all!"*

"Amazing story, Prakash!" said Sally. *"Maybe I will go next. My favorite place to visit is a cafe called Fortitude Coffee in my hometown of Edinburgh, UK. I have not tasted a better avocado toast than they make and I'm looking forward to getting one when I visit later this year,"* she closed.

"It is indeed a real place – I just checked on the internet; £5 for avocado toast is a sweet deal I would say," said Ian from the back of the room. *"Hey, team – I wanted to be part of this exercise since we are going to focus on PTE. Platform dimension which is going to be a critical component of the Resurgence program,"* he said. The other members shared their favorite places and their reasons why over the next 15 minutes. *"I hope that we've now warmed up and are ready for the day. Let me explain what we are going to do next,"* said Prakash. *"We are going to identify all the dimension items – or what we could call topics – that we would like to address under PTE, such as platform dimension. Let's look at the LA.Customer item list that was captured by the Leadership and Architecture stream team yesterday,"* said Prakash while projecting the LA.Customer dimension sheet on the screen. The following figure shows the LA.Customer dimension sheet:

Dimension	Dimension Item	Item Description
LA.Customer	Consistent user interface across all customer applications	Create a consistent UI look and feel across all web and mobile interfaces.
LA.Customer	Cross-channel user experience	Create a seamless cross-channel experience for customers between the web, mobile, IVR, and retail stores. Regardless of which channel they start, provide the ability to leverage other channels as desired to complete the process.
LA.Customer	Personalization	Provide personalization options to the customer so that they can customize the UI as desired across web and mobile channels.
LA.Customer	Self-service	Provide self-service capabilities to make sure customers can sign up, change, pay for, and cancel their services as needed.
LA.Customer	Access	The customer should have at least one channel available all the time – that is, web, mobile app, IVR, or retail store (within 20 miles).
LA.Customer	Data protection	All customer information should be secured in a permanent data store with encryption.
LA.Customer	Customer support	Create multiple channels of support, from FAQ to chat and IVR.

Figure 6.10 – The LA.Customer sheet zoomed in to show the identified items

Prakash then facilitated the 10-for-10 brainstorming and voting exercise, as explained in the previous section, to build and rank the items list for the PTE.Platform dimension. The following figure shows the PTE.Platform sheet, with the items ranked by the number of votes:

Dimension	Dimension Item	Item Description	Percentage Complete	Ranking	Total Votes
PTE.Platform	UnifiedPlatform	A comprehensive and centralized platform that provides a consistent and seamless experience for customers, employees, and partners across all touchpoints and channels	20%	1	12
PTE.Platform	Interoperability	The platform should be able to interact with other systems and applications in the technology ecosystem. It should use open standards and APIs to facilitate integration with other systems, making it easier to share data and functionality across the organization.	20%	2	10
PTE.Platform	Scalability	The platform should be designed to scale horizontally. This can be done by adding more resources (such as servers or databases) to increase capacity as needed.	20%	3	9
PTE.Platform	Flexibility	The platform should be able to evolve and improve over time, incorporating new features and functionality as required.	20%	4	7
PTE.Platform	Security	The platform should incorporate robust authentication and access control mechanisms to protect sensitive data and prevent unauthorized access.	20%	5	6
PTE.Platform	Analytics	The platform should provide analytics data, enabling the organization to track user behavior, measure the effectiveness of the digital initiatives, and make data-driven decisions.	20%	6	5
PTE.Platform	DeveloperCentric	The platforms should be developer-friendly, providing tools and resources that enable developers to build and deploy new workloads and services quickly and easily.	20%	7	4
PTE.Platform	Automation	The platform should utilize automation as much as possible to improve security, availability, and resiliency.	20%	8	4
PTE.Platform	CloudAgnostic	The platform should provide a cloud-agnostic architecture so that applications and workloads can be moved across different providers and between on-prem and cloud environments as required.	20%	9	3
PTE.Platform	Resilient	The platform should be able to withstand any disruptions or challenges that threaten the availability and stability of the platform.	20%	10	3
PTE.Platform	ZeroDowntimePatching	The platform should roll out new patches or upgrades with no downtime to the applications and workloads that are deployed on them. This can be achieved via automation and rolling updates.	20%	11	2
PTE.Platform	EdgeDeployments	The platfrom should be able to provide edge deployment and life cycle management capabilities.	20%	12	1

Figure 6.11 – PTE.Platform sheet with items identified and ranked

With the initial scope defined, Prakash was ready to work with the stream teams to deep dive into the details of those defined items in an operating model dimension.

Detailing dimensions

With the dimensions scoped, the next step was to dive into the details of all the topics (items) that were covered within every single dimension. Prakash guided the stream team and leveraged the template that was provided in this book's GitHub repository as the baseline. The template was to define content, boundaries to other dimensions, and cross-stream collaboration requirements on different aspects of the same dimension. With the collective knowledge spread across the core team and stream teams, along with existing documentation from the enterprise architecture, PMO, security, and compliance teams, the team was able to fill this template across several iterations. They also leveraged industry best practices from different sources as part of this exercise. This was a critical part of the journey since every item that was identified and mapped played an important role in completing the operating model. Like any experienced facilitator, Prakash used several different practices from the OPL to get to the needed level of depth, such as **impact mapping**, **double diamond**, **start at the end**, and **value**

slicing, to determine the transition states. Impact mapping, for example, starts with the goal, which the stream team needed to agree on. In this case, it was fairly simple. The goal was to detail the scope of the `PTE.Platform.UnifiedPlatform` item by defining the target state, start state, and all transition states in between.

Prakash said *"Now, let's detail the PTE.Platform.UnifiedPlatform item by defining the start state, target state, and all the transition states in between. For example, if we want automated CI/CD pipelines for our DevSecOps capability, then perhaps we can start with automated deployments as a first iteration. But that's just an example – I leave that up to you to determine. For PTE.Platform.UnifiedPlatform, let's first define the target state."*

"That's easy," stated Henrik. *"All new workloads and migrated/modernized applications should be deployed to the unified platform. All remaining workloads have a clear modernization or retirement strategy defined and being worked on."*

"Ok, great – do you all agree?" asked Prakash.

"Looks perfect to me," said Gail, looking at the other members.

"Great – if there are no other suggestions, let's finalize this as the target state and now focus on the current state," said Prakash.

The team debated a bit about whether the Java application server they used at Wisent to deploy a few applications together could be considered a unified platform. However, they quickly ruled out that being the case since this environment was about to be retired soon once these applications had been modernized to cloud-native microservices, which was going to happen in the coming months. Courtney mentioned, *"I think we should define the current state as multiple siloed platforms with different technology and operations teams."* Prakash captured it and followed a similar approach to help the steam team define the transition states for the `PTE.Platform.UnifiedPlatform` item. The defined transitions state were as follows:

- **Iteration 1**: Create a unified platform architecture and migration plan to support all modern cloud-native workloads and existing traditional workloads

- **Iteration 2**: Build a unified cloud platform to address all business needs, including faster innovation and standardized operations

- **Iteration 3**: Build and deploy modern cloud-native workloads to the unified platform

- **Iteration 4**: Migrate existing workloads and APIs to the unified platform

- **Iteration 5**: Work on a rebuild/retirement plan for workloads that can be fit into the unified platform

Prakash captured all the state details in the Bison operating model document. The following figure shows the start state, target state, and transition states the steam team defined for the `PTE.Platform.UnifiedPlatform` item:

Iteration 1–Transition State	Iteration 2–Transition State	Iteration 3–Transition State	Iteration 4–Transition State	Iteration 5–Transition State	Target State
Create a unified platform architecture and migration plan to support all modern cloud native workloads and existing traditional workloads.	Build a unified cloud platform to address all business needs including faster innovation and standardized operations.	Build and deploy modern cloud native workloads to the unified platform.	Migrate existing workloads and APIs to the unified platform.	Work on a rebuild/ retirement plan for workloads that can be fit into the unified platform.	All new workloads and migrated/ modernized application deployed to Unified platform. All remaining workloads have clear modernization or retirement strategy defined and worked on.

Figure 6.12 – Transition states defined for PTE.Platform.UnifiedPlatform

> **Note**
>
> This Bison operating model document example is available in this book's GitHub repository (`https://github.com/PacktPublishing/Technology-Operating-Models-for-Cloud-and-Edge`) for you to reference.

Prakash then continued the same exercise for the other items under the `PTE.Platform` dimension and captured them in the Operating Model document, in the `PTE.Platform` sheet:

Start state	Current state	Notes	Iteration 1 - Transition State	Iteration 2 - Transition State
Multiple siloed platforms with different technology and operations teams.	Start state	Across Bison and Wisent, different platforms and technologies are used. They were adopted organically with little standardization work so far.	Create a unified platform architecture and migration plan to support all modern cloud-native workloads and existing traditional work loads.	Build a unified cloud platform to address all business needs, including faster innovation and standardized operations.
Poor interoperability as the data layer is the only option provided for different workloads to interact.	Start state	No clearly defined. A standards-based API layer exists that makes workloads built with different technology stacks interact with one another.	Design API guidelines and architecture for all workloads to follow.	Implement APIs for all web workloads deployed on the unified cloud platform.
Partially implemented; few web workloads have horizontal scalability option enabled.	Start state	Only a few web workloads have this capability. Traditional home-grown or third-party workloads and databases do not have this capability.	Design horizontal scability with auto load balancing as part of the unified cloud platform architecture.	Implement horizontal auto-scaling for web workloads on the unified cloud platform.
Rigid platforms built for workloads of defined characteristics such as technology, framework, architecture, and so on.	Start state	Purpose-built p latforms to support technologies or architecture under consideration.	Define standards for unified cloud platform to be extendable and flexible to support multiple technologies and architecture to address standardization and future business needs.	Implement new web apps to leverage extensibility and flexibility standards defined for the unified cloud platform.
Siloed security implementations for platforms and infrastructure, depending on where they are hosted and what business workloads they support.	Start state	No centralized approach. Both Bison and Wisent have an isolated and localized strategy to address security needs.	Build a standard security architecture and guidelines for a unified cloud platform for supporting all application and data workloads.	Implement new and existing web apps deployed to the unified cloud platform to leverage defined security standards.
No centralized analytics implemented.	Start state	Few modern web apps have monitoring dashboards but not standardized across Bison and Wisent.	Build a standard analytics framework for a unified cloud platform that all workloads can leverage.	Integrate all web workloads on the un ified cloud platform with the newly implemented analytics framework.
No developer portal, self-service capabilities or market place exist.	Start state	Very less focus on developers with multiple different IDEs and tools used by respective development teams that suit the application development they work on.	Build an internal developer platform (IDP) and a rich marketplace to support all development frameworks and tools that are utilized by the developers. Ensure the developer platform is flexible and extendable to support future developer needs.	Ensure all developers are trained on using the IDP and marketplace to support their projects.
Automation exists in pockets. Infrastructure build and app delivery functions are mostly automated, but not enough or consistent to be considered meaningful.	Start state	Infrastructure provisioning in the cloud and virtualized environments are fairly automated. Most also follow CI/CD for app release. However, other functions such as security and compliance and patching are not automated.	Implement a GitOps model for infrastructure, platform, and applicationil development and delivery automation.	Ensure GitOps is implemented for infrastructure provisioning across the edge, on-premise, and cloud data centers.

Figure 6.13 – PTE.Platform state mapping – zoomed-in view

As part of this exercise, the team identified the dependencies and documented them in the template. They also built out the dependencies as and when they identified them, sometimes collaborating with the corresponding stream team as part of their process. For example, the two dependencies that were identified under **LA.Customer** were **LA.Architecture.Resilience** and **LA.Architecture.Security**. The streams team completed the build-out of these two dependencies, as shown here. The following figure shows the **LA.Architecture.Resilience** and **LA.Architecture.Security** dependencies, which were built out as part of the **LA.Customer** dimension build-out process:

Dimension	Dimension Item	Item Description	Related To - Dependent On
LA. Architecture	Resliency	The architecture should be designed to be resilient to support business growth against technical and external disruptions. It should be flexible enough to accommodate new business requirements and also external security and compliance requirements that are forced on the organization.	LA.Customer. Access
LA. Architecture	Scalability	The architecture should be designed to scale as your business grows. It should be able to handle a large volume of traffic and data without slowing down or crashing.	
LA. Architecture	Flexibility	The architecture should be flexible enough to accommodate changes in your business requirements and customer needs. It should allow you to add new features, upgrade technology, and adapt to changing market conditions.	
LA. Architecture	Security	The architecture should prioritize security to protect sensitive data and transactions. This includes measures such as encryption, firewalls, and access controls.	LA.Customer. Data protection

Figure 6.14 – Dependencies built out as part of items definition and tracking (partial view)

The following figure shows the percentages associated with the "definition of done," as highlighted in *Chapter 5*:

Status	Status Definitions	% complete
Not yet started		0%
Discussed	The iterational item has been discussed with the relevant stakeholders	5%
Agreed	The implementation design is agreed upon by all stakeholders.	20%
Designed	The implementation design has been documented and reviewed and approved by the relevant stakeholder groups.	40%
Implemented	The iterational item has been implemented as per the agreed-upon design.	60%
Tested	The design and implementation has been validated according to the stakeholder agreement.	80%
In use	The implementatl on has been tested and is now in use.	90%
Done	All necessary information has been documented and is accessible to all the relevant stakeholders.	100%

Figure 6.15 – Percentage complete status per operating model dimension line item

Since most of the dimension items were in an agreed-up state, they were marked as 20% complete to start with at the end of this stage.

As the next step, after the dimensions were defined and scoped and the dependencies were mapped, the core team populated the program dashboard based on the details that were gathered during the selection, scoping, and detailing exercises across the streams:

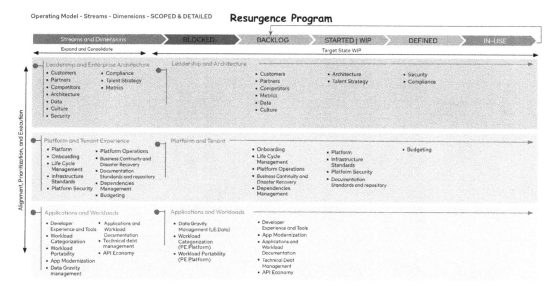

Figure 6.16 – Updated Resurgence program operating model dashboard

Measuring progress

To get a picture of where the operating model was in terms of completeness, the individual line items of the operating model dimension were rolled up into the dimension, and then the dimensions into their respective streams. The overall progress in the Operating Model dashboard showed the progress that had been made.

When program manager Paul was rushing to the car at 4:00 P.M. to pick up his kids from childcare, he bumped into Ian. "*Hey, Paul, how are things? Can I get a report on where the operating model is in terms of completeness by tomorrow morning for the board meeting at 9:00 A.M.? I need a breakdown per stream and an overview of all the work items and their progress.*"

"*Not a problem, Ian – it's already been done. It's on our Operating Model landing page in our wiki. Just follow the link named 'Operating Model Dashboard' – you will get the up-to-date status of the operating model.*"

And even though, in reality, this rarely happens, Ian said, *"Fantastic! That'll do, Paul. Thank you, and have a good evening."*

After this conversation with Paul, Ian went back to his desk and followed the link that Paul mentioned, which brought up the following dashboard:

Stream ID	Categories	Business Relevance and Purpose	# Items	Overall Percentage Complete
1	Leadership and Architecture Stream	The leadership and architecture stream involves establishing clear goals, defining roles and responsibilities, and providing direction and guidance to the team. This includes setting a vision for the business, creating a strategy to achieve it, and managing and motivating the employees to execute that strategy effectively. Effective leadership can help build a strong and resilient organization that can adapt to changing market conditions and emerging opportunities. It also involves designing and implementing the technical infrastructure that supports the digital business. This includes creating a technology roadmap that outlines the systems, tools, and platforms required to meet business objectives, and ensuring that these systems are scalable, secure, and reliable.	66	20.29%
2	Platform and Tenant Experience Stream	The platform and tenant experience stream involves developing and maintaining the technical infrastructure that supports the business, such as your website, mobile app, or other digital platforms and core applications. This includes selecting the appropriate technology stack, managing the software development process, and ensuring that your platforms are scalable, reliable, and secure. An effective platform can help you deliver a seamless user experience, enhance customer engagement, and drive business growth. It also involves creating a positive experience for customers who use your digital platforms or services. This includes designing user interfaces that are intuitive and easy to navigate, providing responsive customer support, and offering personalized experiences based on customer data and preferences. An effective tenant experience can help you build customer loyalty, increase retention rates, and drive revenue growth.	73	22.00%
3	Applications and Workloads	The applications and workloads stream involves developing, purchasing, and maintaining the software applications that power the business. This includes developing custom applications, buying third-party applications from independent software vendors, integrating third-party applications, and ensuring that your applications are scalable, reliable, and secure. Effective applications can help streamline operations, automate processes, and enhance the customer experience. It also involves optimizing the infrastructure that supports your applications and workloads. This includes selecting the appropriate cloud or on-premise infrastructure, adding edge infrastructure as needed, managing capacity and performance, and ensuring that your infrastructure is secure and compliant. Effective workloads can help you reduce costs, improve efficiency, and enhance the user experience for your customers.	58	20.00%
				62.29%
				3
	Overall Operating Model Completeness		197	20.76%

Figure 6.17 – Operating model dashboard with streams progress captured

Similarly, the template also had a dimensions progress dashboard that captured the cumulative progress of all the items that were mapped under the corresponding dimension. The following figure shows the leadership and architecture dimension dashboard:

Categories	Business Relevance and Purpose	#Items	Overall Percentage Complete
Customer	An organization or individual who purchases or uses your products or services. Customers are the primary source of revenue for digital businesses, and building and maintaining positive relationships with them is critical to the success of your business.	7	22.86%
Partner	Another organization or individual with whom you collaborate to achieve common goals or objectives. In the context of a digital business, a partner may be a supplier, a customer, a distributor, a strategically, or even a technology provider.	6	20.00%
Competitor	Organization or individual that offers similar products or services as your business and competes with you for the same target market.	7	20.00%
Architecture	Architecture refers to the design and organization of the digital systems and infrastructure. It includes the various components of the digital ecosystem, such as the software applications, database,s, servers, networks, and other technical systems.	8	20.00%
Data	Data refers to the information that the business collects, processes, and analyzes in the course of its operations. Data can come from various sources, including customer interactions, website analytics, social media, and other digital channels.	8	20.00%
Culture	Culture refers to the shared values, beliefs, and behaviors that guide the actions and decisions of the organization. Culture encompasses the attitudes and approaches that your employees ta ke toward technology, innovation, collaboration, and customer service.	7	20.00%
Security	Security refers to the measures and protocols that are implemented to protect the digital and physical assets, including data, software, hardware, and networks. Security is critical for digital businesses because cyber threats such as data breaches, hacking, and malware attacks can cause significant harm to the business's operations and reputation, as well as customer trust.	8	20.00%
Compliance	Compliance refers to adhering to the various laws, regulations, and industry standards that govern the operations, particularly in the areas of data privacy, security, and transparency. Compliance is important for digital businesses because failure to comply with regulations can result in legal and financial consequences, as well as damage to the business's reputation and customer trust.	5	20.00%
Talent Strategy	Talent strategy refers to the approach you take to attract, develop, and retain the talent you need to succeed in the digital marketplace. Talent strategy is critical for your success as it will help organizations build a team of skilled and motivated professionals who can help you achieve your business goals and stay ahead of their competitors.	5	20.00%
Metrics	Metrics are the quantitative measures used to track and evaluate the performance of the business and its various functions. Metrics allow organizations to understand how the business is performing, identify areas for improvement, and make data-driven decisions to optimize the operations.	5	20.00%
			202.86% 10
Overall Operating Model Completeness		**66**	**20.29%**

Figure 6.18 – Leadership and architecture dashboard

The following figure shows the relationship between the streams, dimensions, and items under the Operating Model:

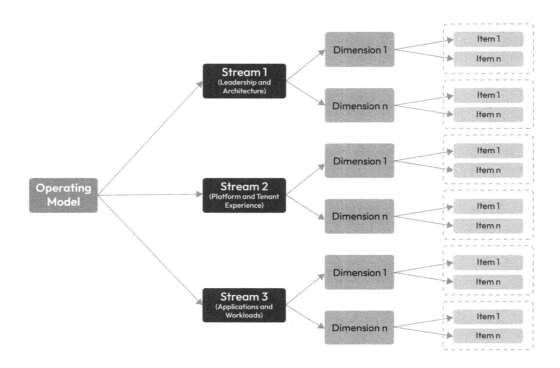

Figure 6.19 – Relationship between streams, dimensions, and items

At this point, it is worth highlighting that the Bison Insurance example is just a guidance based on our experience. Depending on your organization and team structure, you may want to rework the template for a better fit. Some of the key variable factors are around the depth of items to cover under each dimension. A couple of questions I got when I reviewed the template included, "*Does the chatbot qualify to be captured under the leadership and architecture stream?*" and "*Is the competitor a dimension to consider? Shouldn't it be part of our SWOT analysis and strategy work?*" As with most architecture questions, the answer is, "*It depends.*" As you set up your core team and go through the process of workshops and interviews to build the operating model, you may find a certain approach that suits your organization better. You may also find the Bison Insurance case study to be too elaborate with different architecture and business considerations baked in. This is done by design to give you a case study to practice and train on beyond the template that we have built and shared. You may start with the Bison Insurance example template, use the case study, or start from scratch. As part of this exercise, you will quickly realize the importance of addressing all the streams before building the platform and operating model. It is easy and sometimes logical to work on the streams independently at different times. Due to the organic nature of many cloud journeys, we have seen organizations build the platform first and try to fit other dimensions on it and extend the platform only when it is no longer sufficient. This approach may have worked historically when technology was a supporting function or part enabler, but now, with technology being vital to survive and thrive, this is no longer the case.

Another important thing we want to highlight is that the Resurgence program funding, team setup, and proposed execution model that we have explained in this chapter (and will continue doing so in the next) is not the standard approach we would recommend for all organizations. Depending on how your organization is set up and where you are on your cloud and edge journey, your approach may differ. You may not have the luxury of building a ground-up program with 5 years of funding secured like Bison Insurance. On the other hand, you may also have a setup where the entire organization is heavily invested in the program, with stream members allocated full time for the first phase of the initiative, which is building the Operating Model and implementing it. Depending on your situation, you may start at different places and adopt a different approach that suits your needs. The key requirement is to have an accountable core team and executive sponsorship at the CXO level.

Summary

In this chapter, we introduced a fictitious case study of an insurance company called Bison and looked at its growth, acquisitions, competitive threats, global expansion plans, and digital transformation initiative. We showed how Bison Insurance created a new program called Resurgence program and built a core team to lead this initiative.

We also showed how Bison Insurance built their distributed cloud and edge operating model while following the template and open principles that were explained in *Chapter 5*, *Building Your Distributed Technology Operating Model*.

The Bison Insurance operating model documentation is available in this book's GitHub repository for reference: `https://github.com/PacktPublishing/Technology-Operating-Models-for-Cloud-and-Edge`.

In the next chapter, we will learn how enterprise-grade open source technologies were leveraged to build the Resurgence platform, a distributed cloud and edge platform based on their distributed cloud and edge operating model.

Further reading

To learn more about the topics covered in this chapter, take a look at the following icebreaker activities: `https://www.atlassian.com/team-playbook/plays/icebreaker-activities`.

7

Implementing Distributed Cloud and Edge Platforms with Enterprise Open Source Technologies

In the previous chapter, we learned how Bison Insurance built its distributed cloud and edge operating model following the processes and template defined in *Chapter 5, Building Your Distributed Technology Operating Model*. We followed six main steps to determine our operating model dimensions for each operating model stream and, subsequently, our complete technology operating model. As in *Chapter 5*, we defined streams and grouped dimensions under each stream that the streams own, and subsequently built out an operating model for the resurgence program. In this chapter, we will show how Bison Insurance built its distributed cloud and edge platform called the Resurgence platform, leveraging enterprise open source technologies. At the end of this chapter, we will have achieved the following:

- Demonstrated how enterprise-grade open source technologies have become the core innovation engine for the implementation of a distributed cloud and edge architecture

- Learned how Ian, Donna, and the core team architected the Resurgence platform in alignment with their technology operating model

The following topics will be covered in this chapter:

- Open source as the innovation engine

- Building distributed cloud and edge platforms with enterprise open source technologies as the foundation

Let us start this chapter, *Implementing Distributed Cloud and Edge Platforms with Enterprise Open Source Technologies*, from where we left off in *Chapter 6, Your Distributed Technology Operating Model in Action*.

Building a Distributed Cloud and Edge Platform

Once the operating model was defined, the core team worked on building the components and processes to implement the operating model. For the purpose of this book, let us take the *Platform and Tenant Experience* stream and explain how Bison Insurance built its distributed cloud and edge platform covering all dimensions, items, and their dependencies that are captured in the Bison operating model. The following screenshot shows the Platform and Tenant Experience stream dashboard:

Categories	Business Relevance & Purpose	# Items	Overall Percentage Complete
Platform	A platform refers to a digital infrastructure or ecosystem that allows you to connect and engage with the customers, partners, and other stakeholders. A platform typically provides a set of tools, services, and interfaces that enable users to interact with organizations business and each other.	11	40.00%
Onboarding	Onboarding refers to the process of integrating new customers or employees into organization's ecosystem. It typically involves a series of steps and activities aimed at familiarizing the new user with organization's products or services, as well as the business culture, processes, and policies.	8	20.00%
Life Cycle Management	Platform lifecycle management refers to the process of managing the different stages of a digital platform, from conception to retirement. It involves planning, building, launching, maintaining, and retiring a platform in a systematic and strategic manner, in order to optimize its performance, usability, and longevity.	8	20.00%
Infrastructure standards	Infrastructure standards refer to the guidelines and best practices for designing, building, and maintaining the physical and digital infrastructure that supports organization's digital operations. This includes hardware, software, networking, and cloud infrastructure, as well as data centers, servers, and other physical components.	10	20.00%
Platform Security	Platform security refers to the measures and protocols that are put in place to protect the digital platform, as well as the data and information that are stored on it, from unauthorized access, theft, or loss. Platform security is a critical aspect of digital business operations, as it helps safeguard the business and its customers from potential cyberattacks and other security threats.	7	20.00%
Platform Operations	Platform operations refer to the management and maintenance of the digital platform that supports the business operations. This includes the hardware, software, and network infrastructure that underpins the platform, as well as the processes and procedures that are necessary to ensure its ongoing performance, reliability, and scalability.	6	20.00%
Business Continuity & DR	Business continuity and disaster recovery refer to the processes and procedures that are put in place to ensure the ongoing availability and resilience of the digital platform, data, and operations, in the event of a disruptive event or disaster.	5	20.00%
Documentation standards	Documentation standards refer to the guidelines and best practices that are followed for creating, organizing, and maintaining documentation related to the digital platform, products, services, and processes. Documentation is an essential part of running a digital business, as it helps ensure consistency, quality, and efficiency in your operations, and facilitates collaboration and communication among team members and stakeholders.	8	20.00%
Dependency Management	Dependency management refers to the process of identifying, tracking, and managing the various software and infrastructure dependencies that the platform and products rely on. This includes third-party libraries, APIs, frameworks, and other components that are used to build, deploy, and run all digital applications.	5	20.00%
Budgeting	Platform budgeting refers to the process of planning, allocating, and managing financial resources for your digital platform and related products and services. This involves identifying the costs and expenses associated with developing, operating, and maintaining your digital platform, and making strategic decisions about how to allocate resources to meet the business objectives and goals.	5	20.00%
			220.00%
			10
Overall Operating Model Completeness		**73**	**22.00%**

Figure 7.1 – Platform and Tenant Experience stream dashboard

One of the key dimensions of this stream is the platform dimension. Most other dimensions that are listed under this stream are built upon the platform. The following screenshot shows the `PTE.Platform` dimension dashboard, where all items under the platform dimension are captured:

Dimension	Dimension Item	Item Description
PTE.Platform	UnifiedPlatform	A comprehensive and centralized platform that provides a consistent and seamless experience for customers, employees, and partners across all touchpoints and channels.
PTE.Platform	Interoperability	Platform should be able to interact with other systems and applications in the technology ecosystem. It should use open standards and APIs to facilitate integration with other systems, making it easier to share data and functionality across the organization.
PTE.Platform	Scalability	Platform should be designed to scale horizontally by adding more resources (such as servers or databases) to increase capacity as needed.
PTE.Platform	Flexibility	Platform should be able to evolve and improve over time, incorporating new features and functionality as required.
PTE.Platform	Security	Platform should incorporate robust authentication and access control mechanisms to protect sensitive data and prevent unauthorized access.
PTE.Platform	Analytics	Platform should provide analytics enabling organization to track user behavior, measure the effectiveness of the digital initiatives, and make data-driven decisions.
PTE.Platform	DeveloperCentric	Platforms should be developer-friendly, providing tools and resources that enable developers to build and deploy new workloads and services quickly and easily.

Figure 7.2 – PTE.Platform dashboard with all items defined (partial view)

With the items defined and dependencies mapped, the core team started working on building a logical architecture for the future platform, as explained in the following sections.

Platform and Platform team

At this point, it is important to revisit the terms *platform*, *platform team*, and *Platform as a Product*, which we explained in *Chapter 1, Fundamentals of the Cloud Operating Model*. A **platform** is an integrated collection of features and functions defined and presented according to the needs of the users; in our context, it is the developers, operators, and security teams. The platform consists of self-service APIs, tools, services, knowledge management features, and integrated support. The platform is managed as a compelling internal product, is accessible via a web browser for ease of access, and is used to help drive standardization. Delivery teams can make use of the platform to deliver product features faster and reduce cognitive load and coordination efforts.

The most effective platforms are managed by a **platform team**. The goal of the platform team is to improve developer productivity and operational efficiency by abstracting the complexity of distributed cloud and edge environments away from development teams by providing a curated set of standardized APIs. This helps the teams leverage the platform to integrate/plug in these curated services with clearly defined **Service Level Agreements (SLAs)**, letting them focus on their primary responsibilities, which are directly aligned with delivering faster and better business outcomes.

A detailed discussion about how to build and manage a platform team is beyond the scope of this book. However, I strongly recommend you spend time understanding and implementing a platform team to build and manage your platform. One key aspect to focus on as part of this exercise is the **Platform-as-a-Product** model. Building the platform as a product differs from running projects with defined start and end dates. Platforms need continuous iterative product development with ongoing improvements to enhance the **user experience** (**UX**) and the business value they deliver. The platform team should be constantly receiving feedback from business and technical stakeholders in order to enrich their release planning and regularly ship new features that are fit for purpose, along with ensuring the resilience and usability of the platform. The following diagram shows the core components of a modern cloud platform:

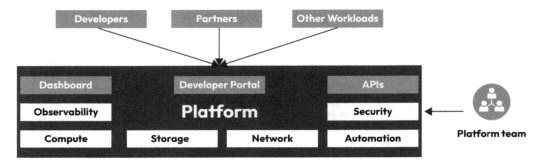

Figure 7.3 – Key components of the platform

After some analysis of the operating model requirements across all streams and management discussions, the core team has decided to build a platform team and operate it as a product. They call the new platform the **Resurgence platform** and the platform team the **Resurgence platform team**. We will see how they build the platform in the remaining sections.

Implementing the platform based on the operating model

Based on the key requirements captured in the operating model, the core team started with building a logical architecture with all critical elements of the Resurgence platform. They focused on highlighting attributes that are essential such as cloud agnostic, APIs, developer portal, and so on, which are captured under the PTE.Platform dimension. This ensures that their platform is not only able to host workloads but also able to address developer tools and partner integrations to connect with legacy applications that will not be hosted on the platform. They also highlighted the need for automation, security, and compliance, which covers all aspects of the platform, as captured in the operating model. The following diagram shows the proposed logical architecture of the Resurgence platform:

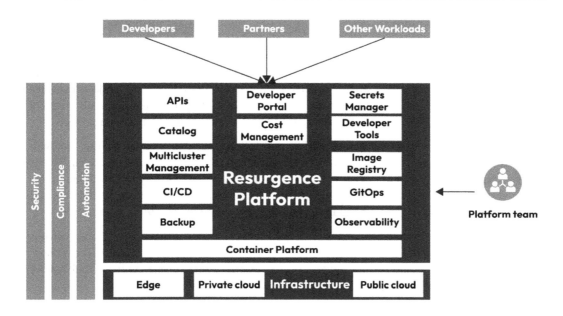

Figure 7.4 – Resurgence program platform architecture

In the next section, we will focus on how the core team leveraged enterprise open source technologies in building the Resurgence platform.

The power of enterprise open source

With the logical architecture in place, the team agreed upon building the Resurgence platform on enterprise open source technologies. Given that open source projects have become the innovation standard, adopting enterprise open source technologies is a sensible decision to make. They created guidelines for selecting vendors to support the Resurgence platform and got it reviewed and agreed on by the Platform and Tenant Experience stream team. (It may be a good idea to run this exercise as a workshop if you want a stream team to drive this end to end.) The following are the guiding principles that are created for selecting enterprise open source technologies for the Resurgence platform utilizing the MoSCoW framework:

- **Open source expertise**: The selected vendor must have a proven history of providing enterprise-grade open source software solutions that are essential for building the Resurgence platform, from operating systems to container orchestration technologies such as Kubernetes and container runtime environments such as CRI-O. Their expertise in these areas helps ensure that the hybrid cloud and edge environment is stable, secure, supported, and evolving alongside the open source community.

- **Open source commitment**: The vendor must be committed to the open source community, which means that their software is freely available and can be customized to meet an enterprise's specific needs. This openness can be beneficial for enterprises that want to avoid vendor lock-in and maintain sovereign control over their cloud infrastructure. Furthermore, the vendor should offer co-engineering relationships with their customers to allow open source projects to evolve in a specific direction by adding the functionality needed for their customers. This also reduces technical debt by contributing this functionality to the corresponding upstream projects.

- **Open standards stack**: The vendor's software solutions must be designed to integrate with existing processes and systems and must be open standards-based and API driven. This is to simplify the process of building and managing components of a hybrid cloud and edge environment. This integration can help us avoid compatibility issues and reduce the time and effort required to set up and maintain tools, processes, and infrastructure.

- **Support**: The vendor must offer enterprise-level support for their software solutions, which is important for enterprises that need timely assistance in the event of technical issues. The vendor's support team must be staffed by **subject-matter experts** (**SMEs**) who can help troubleshoot problems and provide guidance on best practices.

- **Innovation focus**: The vendor must be acknowledged for their commitment to innovation by the open source community and the **Cloud Native Computing Foundation** (**CNCF**), and continue to fund new technologies and tools that can help enterprises build and manage their distributed infrastructure more effectively. By working with the vendor, organizations must be able to stay up to date with the latest developments in the open source community and take advantage of new developments as they emerge.

- **Services and partner network**: The vendor must be able to help customers implement open source solutions with their globally distributed services team and partner network. A partner training and certification mechanism must be in place.

Paul presented the reviewed and approved enterprise open source platform guiding principles to Donna with detailed documentation. Donna liked the proposal but wanted to get more details on how the platform would support their existing applications and workloads that are not ready to be deployed to the platform. She wanted Paul to work with the core team to understand how existing applications can be migrated to the Resurgence platform.

Prakash and Liang worked on an application migration and modernization strategy for all applications to be either migrated or modernized to the platform or retired in the next 3 years. They leveraged the Konveyor project (`https://www.konveyor.io/`) under CNCF to help them with this initiative. Konveyor provided automated application analysis for existing (Java and Node) applications and gave them a score to help Bison Insurance categorize the application based on the technical complexity into different buckets, such as *Rehost, Replatform, or Refactor*, which are represented in the 6R application migration framework introduced by Amazon (https://aws.amazon.com/blogs/enterprise-strategy/6-strategies-for-migrating-applications-to-the-cloud/). With this categorization and the application migration intelligence gathered from Konveyor tools, the team was able to come up with a detailed application categorization and migration plan. They reviewed the application categorization and migration plan with the Platform and Tenant Experience and the Applications and Workloads stream teams to ensure that the application categorization was aligned with the Resurgence program objectives. Upon agreement with the stream teams, Prakash and Liang presented the finalized applications' categorization and migration plan to Donna. Donna reviewed the plan and liked the detailed work that had been put together by the team. Donna consulted with Ian, and Ian suggested setting up a meeting with the core team so that they could synchronize across the topics of platform architecture, application categorization, and migration plan. The meeting was scheduled for Friday. Paul asked Prakash and Liang to lead the presentation with Miguel's and Susan's support.

At the Friday meeting, the core team led by Prakash and Liang explained the overall platform architecture as well as the application migration and modernization strategy. Miguel and Susan pitched in with technical details as and when Ian wanted to learn more. Ian was particularly interested in understanding how the new cloud-native applications will be able to connect and extend to traditional applications that are not on the Resurgence platform initially and later be moved onto the platform or retired. Susan explained the research she had done on this topic and how enterprise-grade open source products, tools, and API management capabilities that are part of the Resurgence platform architecture can address this challenge. Ian appreciated the thorough work the core team had put together and said that all looked great; however, he said he would still like to see a detailed architecture diagram, including a mapping against the platform capabilities, in their next monthly meeting to ensure they had covered all aspects of the operating model. He reminded Donna of the importance of security and compliance and a flexible API layer for partner integration. Donna said to Ian, "*I am positive we will be able to present this at the next monthly meeting and address all your requests and concerns and also get the proposed architecture reviewed and approved by the stream team.*" The next section explains how the core team mapped the platform capabilities to build the Resurgence platform.

Mapping platform capabilities

Paul realized that the next monthly meeting was only 2 weeks away and there was a long weekend in between. Paul sent a note to the core team on their Slack channel and asked Susan to lead this workstream. Donna also wanted to participate in the workstream so that she could validate the application migration and modernization architecture with her vast experience managing Wisent's applications.

Susan then worked on building the platform capabilities map to address all key capabilities that needed to be included in the platform. The following diagram shows the Resurgence platform capabilities diagram that Susan built:

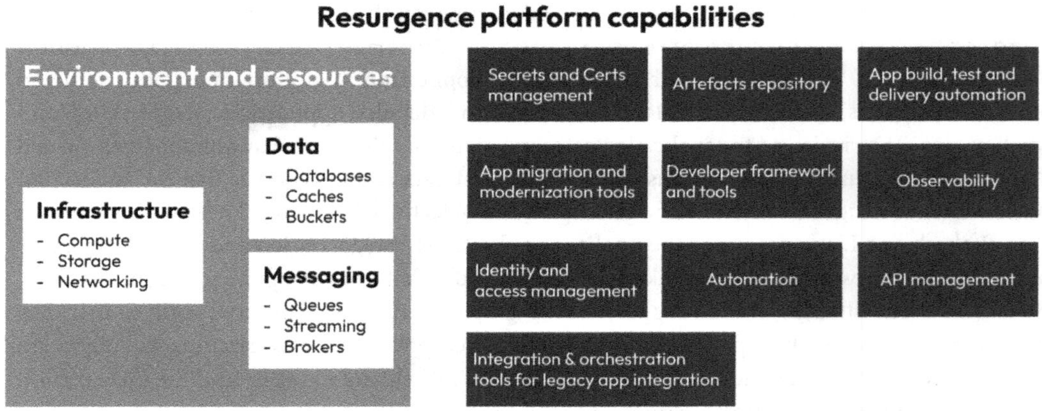

Figure 7.5 – Resurgence platform capabilities

> **Note**
>
> Even though the term cluster has been around for a while, in the hyperscaler context it is quite important to differentiate precisely how we describe cloud environments. Environments built out of proprietary hyperscaler services are hard to reuse elsewhere. In order to achieve true hyperscaler independence, we use the term cluster as the standard environment deployment unit of choice. Clusters can be built, resized, and rebuilt anywhere and offer consistent APIs and UX, irrespective of the infrastructure used. We will be using the term *multicluster* in this book to reflect multiple cloud environments.

The team reviewed the capabilities diagram and agreed that it covered most key aspects to address all items under the **PTE.Architecture** dimension. Donna was quick to note that the current diagram seemed to miss multicluster management, policy management, and enforcement, as well as edge capabilities. Susan corrected it and sent the revised version to the team. The following diagram shows the revised version with edge deployments included:

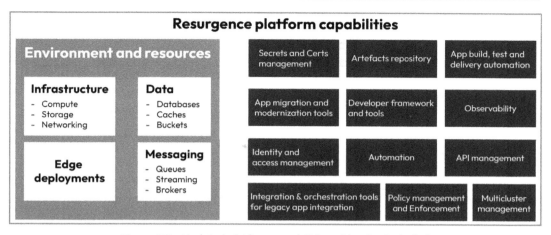

Figure 7.6 – Updated platform capabilities with edge included

Once the core team reviewed and agreed on the platform capabilities diagram, the team moved on to building the detailed implementation architecture, as explained in the next section.

Building the Resurgence platform detailed architecture

Prakash took the lead on these remaining work items and worked with Susan, and they got back to the team with their findings. They started with identifying the top three enterprise open source products for each required platform capability with additional details. They leveraged a combination of Resurgence platform guiding principles listed previously, internal expertise, Gartner and IDC reports, CNCF landscape documentation, and public customer references to aid them through this process. They captured their findings in the ResurgencePlatformTechnologiesMapping-Choices document, as shown in the following table:

Capability	Product	Backing open source project	Enterprise support vendor	GitHub repository
Edge computing	Red Hat Device Edge	MicroShift, Fedora IoT	Red Hat	`https://github. com/redhat-et/ microshift`
	k3s	k3s	SUSE (Rancher)	`https://github. com/k3s-io/k3s`
	MicroK8s	MicroK8s	Canonical	`https://github. com/canonical/ microk8s`

Java application server	JBoss EAP	WildFly (JBoss)	Red Hat	`https://github.com/wildfly/wildfly`
	WebSphere Liberty	Open Liberty	IBM	`https://github.com/OpenLiberty/open-liberty`
	Apache Geronimo	Apache Geronimo	IBM (Withdrawn)	`https://github.com/apache/geronimo`
Operating systems	Linux (**Red Hat Enterprise Linux (RHEL)**)	Fedora	Red Hat	`https://github.com/topics/fedora-project`
	Linux (**SUSE Linux Enterprise Server (SLES)**)	openSUSE	SUSE	`https://github.com/opensuse`
	Ubuntu Server	Ubuntu	Canonical	`https://github.com/ubuntu`
Observability	Sysdig Secure	Sysdig	Sysdig	`https://github.com/draios/sysdig`
	Thanos	Thanos	<<Unknown>>	`https://github.com/thanos-io/thanos`
	Grafana Enterprise	Prometheus	Grafana Labs	`https://github.com/prometheus/prometheus`
Container registry	Red Hat OpenShift Platform Plus	Quay	Red Hat	`https://github.com/quay/quay`
	VMware Harbor Registry	Harbor	VMware	`https://github.com/goharbor/harbor`
	Dragonfly	Dragonfly	<<Unknown>>	`https://github.com/dragonflyoss/Dragonfly2`

Automation and configuration	Ansible Automation Platform	Ansible	Red Hat	`https://github.com/ansible/ansible`
	Terraform Enterprise	Terraform	HashiCorp	`https://github.com/hashicorp/terraform`
	Puppet Enterprise	Puppet	Puppet	`https://github.com/puppetlabs/puppet`
Keys and certificate management	Vault Enterprise	Vault	Terraform	`https://github.com/hashicorp/vault`
	CyberArk Secrets Hub	CyberArk	Conjur	`https://github.com/cyberark/Secret-Manager-formerly-AAM`
	Teleport Enterprise	Teleport	Teleport	`https://github.com/gravitational`
Integration and orchestration	Red Hat Application Foundations	Camel (and Camel K)	Red Hat	`https://github.com/apache/camel`
	Mule ESB	Mule ESB	Salesforce (MuleSoft)	`https://github.com/mulesoft/mule`
	WSO2	WSO2	WSO2	`https://github.com/wso2-attic/product-esb`
Service mesh	Red Hat OpenShift	Istio	Red Hat	`https://github.com/istio/istio`
	Linkerd	Linkerd	Buoyant	`https://github.com/linkerd/linkerd2`
	Consul Enterprise	Consul	Terraform	`https://github.com/hashicorp/consul`

API gateway	Red Hat Application Foundations	3scale	Red Hat	`https://github.com/3scale/apicast`
	Kong Enterprise	Kong	Kong	`https://github.com/Kong/kong`
	MuleSoft Anypoint Platform	MuleSoft	Salesforce (MuleSoft)	`https://github.com/mulesoft/mule`
Certified Kubernetes distribution	Red Hat OpenShift	OKD	Red Hat	`https://github.com/okd-project/okd`
	Rancher Kubernetes Engine (RKE)	Rancher	SUSE	`https://github.com/rancher/rancher`
	Amazon **EKS Distro (EKS-D)**	**Amazon Web Services (AWS)**	Amazon	`https://github.com/aws/eks-distro`
Microservices framework	Red Hat OpenShift	Quarkus	Red Hat	`https://github.com/quarkusio/quarkus`
	VMware Spring Runtime	Spring Boot	Vmware	`https://github.com/spring-projects/spring-boot`
	Micronaut Premium	Micronaut	Micronaut	`https://github.com/micronaut-projects/micronaut-core`
Security and compliance	Red Hat OpenShift Platform Plus (Advanced Cluster Security)	StackRox	Red Hat	`https://github.com/stackrox/`
	Red Hat OpenShift Platform Plus	Clair	Red Hat	`https://github.com/quay/clair`
	Snyk Enterprise	Snyk	Snyk	`https://github.com/snyk/`

Streaming and messaging	Confluent Platform	Kafka	Confluent	`https://github.com/apache/kafka`
	Luna streaming	Pulsar	DataStax	`https://github.com/apache/pulsar`
	Red Hat Application Foundations	ActiveMQ	Multiple vendors including Red Hat under AMQ support	`https://github.com/apache/activemq`
Database – relational	MySQL Enterprise Edition	MySQL	Oracle	`https://github.com/mysql/mysql-server`
	Crunchy Postgres	PostgreSQL	Crunchy Data	`https://github.com/postgres/postgres`
	MariaDB Enterprise	MariaDB	MariaDB	`https://github.com/MariaDB/server`
Database – NoSQL	DataStax Enterprise	Cassandra	DataStax	`https://github.com/apache/cassandra`
	MongoDB Enterprise	MongoDB	MongoDB	`https://github.com/mongodb/mongo`
	ScyllaDB Enterprise	ScyllaDB	Scylla	`github.com/scylladb/scylla`
Database – distributed	CockroachDB self-hosted	CockroachDB	Cockroach Labs	`https://github.com/cockroachdb/cockroach`
	YugabyteDB Anywhere	YugabyteDB	Yugabyte	`https://github.com/YugaByte/yugabyte-db`
	Redis Enterprise	Redis	Redis	`https://github.com/redis/redis`

CI/CD	Red Hat OpenShift	Argo	Red Hat	https://github.com/argoproj/argo-cd
	Red Hat OpenShift	Tekton	Red Hat	https://github.com/tektoncd/pipeline
	Flux CD	Flux CD	Weaveworks	https://github.com/fluxcd/flux2
Multicluster management	Red Hat Advanced Cluster Management	Open Cluster Management	Red Hat	https://github.com/open-cluster-management-io
	Rancher	Rancher	SUSE	https://github.com/rancher/rancher
Identity and access management (IAM)	Pomerium	Pomerium	Pomerium	https://github.com/pomerium/pomerium
	Red Hat OpenShift	Keycloak	Red Hat	https://github.com/keycloak/keycloak

Table 7.1 – Capability to open source project and project mapping table from the ResurgencePlatformTechnologiesMapping-Choices document

Once the table was completed, the core team ran a workshop with the Platform and Tenant Experience stream team to identify the top enterprise open source technologies per capability. This involved taking one capability at a time and analyzing the enterprise open source products identified across different vectors such as customer adoption, the health of the open source project (for example, community participation), the number of contributors from different organizations, popularity, and partner ecosystem strength to pick the enterprise open source technologies of choice. However, for certain capabilities, they realized that they may need more than one software to address the Resurgence platform requirements and therefore also explored this. Once done, Prakash and Susan captured the decision in the `ResurgencePlatformTechnologiesMapping-Finalized` document, as shown in the following table:

Capability	Product	Backing open source project	Enterprise support vendor	GitHub repository

Edge platform	Red Hat Device Edge	MicroShift, Fedora IoT	Red Hat	`https://github.com/ redhat-et/microshift`
				
				`https://github.com/ fedora-iot`

Java app server	Red Hat Application Foundations (JBoss EAP)	WildFly (JBoss)	Red Hat	`https://github.com/ wildfly/wildfly`
Operating system	**Red Hat Enterprise Linux (RHEL)**	Fedora	Red Hat	`https://github.com/ topics/fedora-project`
Observability	Grafana Enterprise	Prometheus	Grafana Labs	`https://github.com/ prometheus/prometheus`
Container registry	Red Hat OpenShift Platform Plus	Quay	Red Hat	`https://github.com/quay/ quay`
Automation and configuration	Red Hat Ansible Automation Platform	Ansible	Red Hat	`https://github.com/ ansible/ansible`
	Terraform Enterprise	Terraform	HashiCorp	`https://github.com/ hashicorp/terraform`
Keys and certificate management	Vault Enterprise	Vault	Terraform	`https://github.com/ hashicorp/vault`
Integration and orchestration	Red Hat Application Foundations	Camel (and Camel K)	Red Hat	`https://github.com/ apache/camel`
Service mesh	Red Hat OpenShift	Istio	Red Hat	`https://github.com/ istio/istio`
API gateway	Kong Enterprise	Kong	Kong	`https://github.com/Kong/ kong`
	Red Hat Application Foundations	3scale	Red Hat	`https://github. com/3scale/apicast`

Certified Kubernetes distribution	Red Hat OpenShift	OKD	Red Hat	`https://github.com/` `okd-project/okd`
Microservices framework	Red Hat OpenShift	Quarkus	Red Hat	`https://github.com/` `quarkusio/quarkus`
	VMware Spring Runtime	Spring Boot	Vmware	`https://github.com/` `spring-projects/` `spring-boot`
Security and compliance	Red Hat OpenShift Platform Plus (Red Hat Advanced Cluster Security)	StackRox	Red Hat	`https://github.com/` `stackrox/`
	Red Hat OpenShift Platform Plus	Clair	Red Hat	`https://github.com/` `quay/clair`
	Snyk Enterprise	Snyk	Snyk	`https://github.com/` `snyk/`
Streaming and messaging	Confluent Platform	Kafka	Confluent	`https://github.com/` `apache/kafka`
Database – relational	Crunchy Postgres	PostgreSQL	Crunchy Data	`https://github.com/` `postgres/postgres`
Database – NoSQL	DataStax Enterprise	Cassandra	DataStax	`https://github.com/` `apache/cassandra`
Database – distributed	YugabyteDB Anywhere	YugabyteDB	Yugabyte	`https://github.com/` `YugaByte/yugabyte-db`
	Redis Enterprise	Redis	Redis	`https://github.com/` `redis/redis`
CI/CD	Red Hat OpenShift	Argo	Red Hat	`https://github.com/` `argoproj/argo-cd`
	Red Hat OpenShift	Tekton	Red Hat	`https://github.com/` `tektoncd/pipeline`

Multicluster management	Red Hat OpenShift Platform Plus (Red Hat Advanced Cluster Management)	Open Cluster Management	Red Hat	`https://github.com/open-cluster-management-io`
Identity and access management (IAM)	Red Hat OpenShift	Keycloak	Red Hat	`https://github.com/keycloak/keycloak`

Table 7.2 – Capability to open source project and project mapping table from the ResurgencePlatformTechnologiesMapping-Finalized document

> Note
>
> The preceding example is for reference purposes only. Please build your own capabilities to product mapping with your stream teams that align with your platform objectives.

Susan shared the finalized capabilities mapping document on the group channel. With the Resurgence platform capabilities documented and the corresponding enterprise open source technologies identified, Donna wanted to build the Resurgence platform capabilities diagram with mapped enterprise open source technologies. She worked with the core team over a lunch meeting to complete this mapping. The following diagram shows the Resurgence platform capabilities mapped to enterprise-grade open source technologies:

Resurgence platform capabilities - Enterprise open source projects (mapped v1.0)

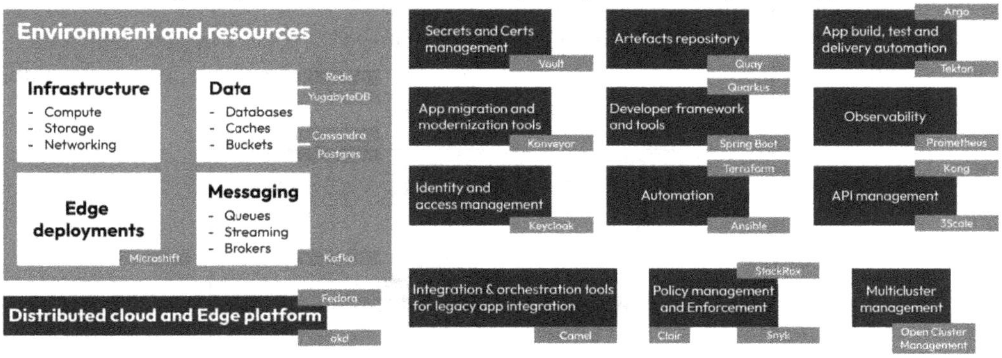

Figure 7.7 – Resurgence platform capabilities mapped to enterprise-grade open source technologies

Donna sent this mapping to Ian for his review, with a clear context on how they selected the products and the reasoning behind leveraging additional technologies from the open source ecosystem. She texted Ian and asked if he had a few minutes for her to walk him through the mapping. Ian responded, *"Please drop by. I have some time now."* Donna walked to Ian's office and showed the mapping that she had printed out on the way. Ian liked the overall mapping but wanted to understand where the public cloud fitted in. Donna explained that it is grouped under infrastructure since the Resurgence platform is cloud agnostic and will support public cloud, private cloud, and edge deployments.

Ian looked at the diagram closely and asked, *"What are our options with regard to outsourcing the operations of our cloud and edge platforms?"* Donna paused and said, *"Great question. We are already utilizing architecture best practices, end-to-end automation, and SRE principles through our guiding principles, but if I heard you correctly, you are thinking about how we can further reduce our running efforts to help us focus on developing differentiating features for our partners and customers. Is that correct?"* Ian said, *"Spot on. I wonder if we can incorporate some managed services offering to the Resurgence platform to address this, either through the vendors directly or their partner ecosystem."* Donna thought for a while and said, *"I think it is going to be difficult to find multiple public cloud-managed services that are fully compatible with the Resurgence platform. I assume building that compatibility and integration across multiple public cloud providers will consume significantly more effort and time that we have not planned and budgeted for"*, which Ian acknowledged. After a few seconds of silence, Donna said, *"Let me go back to my team and get back to you with some options to address this."* Ian agreed.

Donna ended the meeting and immediately posted this question to the team on the team channel from her phone as she walked back to her desk. As soon as she posted the question, she saw a *"Susan is typing..."* message in the channel notification section. After a short while, Susan's response followed. It read: *"Interesting. In fact, we did identify some options and even selected the offering of choice; however, we later dropped it since managed services did not crystalize clearly out of our Leadership and Architecture stream meetings, and finding partners of the required standard would have added additional time and complexity. Let me spend some time on this and get back to you with the managed services platform of choice."* Susan then looked at the earlier version of the `ResurgencePlatformTechnologies Mapping-Choices` document and found the following managed services mapping, which didn't make it to the final list. The following table shows the platform-managed services mapping that the team discussed earlier:

Capability	Open source project	Enterprise support vendor	GitHub repository
Certified application platform – third-party-hosted and managed	Amazon **Elastic Kubernetes Service (EKS)**	Amazon	N/A
	Red Hat OpenShift on AWS (ROSA), Azure Red Hat OpenShift (ARO)	Red Hat	N/A
	Azure Kubernetes Service (AKS)	Microsoft	N/A

Table 7.3 – Managed services choices

Susan sent the following note to the Platform and Tenant Experience stream team in the stream channel: "*Howdy, team. Guess what? The managed services platform capability that we discussed during the last workshop is back on the table. We would like to consider this capability because it allows us to better focus on CX and differentiating features. I wanted to get your agreement one more time on the managed services platform we selected during the workshop. For your reference, please find below the options we discussed:*

Capability	Open source project	Enterprise support vendor	GitHub repository
Certified application platform – third-party-hosted and managed	Amazon **Elastic Kubernetes Service (EKS)**	Amazon	N/A
	Red Hat OpenShift on AWS (ROSA), Azure Red Hat OpenShift (ARO)	Red Hat	N/A
	Azure Kubernetes Service (AKS)	Microsoft	N/A

Figure 7.8 – Identified hosted application platform choices

We chose Red Hat Cloud Services since it aligns closely with the Resurgence platform requirements. Can you send your decision by end of day today so that we can finalize this and move it to the Resurgenc ePlatformTechnologiesMapping-Finalized *document? In case we do not get the majority vote required for the approval, let us run a workshop to get this properly discussed and agreed.*" The stream team quickly got back with their approval since this discussion was still fresh in their mind.

With the required approval in place, Susan then updated the ResurgencePlatformTechno logiesMapping-Finalized document and sent the following note to the core team in the team channel: "*Team, we got agreement. It is Red Hat Cloud Services.*" Susan continued that Red Hat provides the enterprise-supported version of OKD—which is the agreed "distributed cloud and edge platform" for their Resurgence platform—as native cloud services in both **Amazon Web Services (AWS)** and Microsoft Azure. Those services can be integrated with the Resurgence platform. The Platform and Tenant Experience stream team selected this as the preferred choice since it aligns best with the Resurgence platform guiding principles compared to the other choices. She also shared the website link to Red Hat Cloud Services in the channel. Donna walked to Susan's desk and asked, "*Do you have a few minutes to discuss this?*", pointing to the message Susan had shared about Red Hat Cloud Services. "*Happy to,*" replied Susan and explained to Donna how Red Hat Cloud Services could be added to the Resurgence platform. Donna spent a few minutes on the documentation and walked back to Ian's office with excitement. Ian was surprised to see her back so soon and said, "*Don't tell me you have solved my problem already?*" with a smile. "*You bet. Our team did,*" replied Donna and showed the Red Hat Cloud Services website to Ian and walked him through the details of **Red Hat OpenShift on AWS (ROSA)** and **Azure Red Hat OpenShift (ARO)**. Ian was happy to hear about this and asked Donna to incorporate it into the mapping diagram and share it with him. Donna mentioned she would have the document uploaded to the wiki and would share the URL with him in the next hour.

Donna updated the mapping diagram shared with the core team for review. The following diagram shows the Red Hat platform capabilities mapping with Red Hat Cloud Services included:

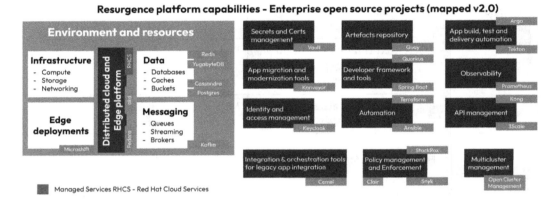

Figure 7.9 – Updated Resurgence platform mapping with Red Hat Cloud Services

Upon review, Donna uploaded the diagram to their internal wiki along with other documents from this workstream and shared the URL with Ian. Before Ian could check the updates and respond, Prakash remembered that they had missed adding a key element to the platform capabilities: the **Internal Developer Platform (IDP)**. Ian had mentioned it should be included in the first version of the platform during the last monthly meeting; he even shared the latest article about how Backstage has helped Spotify to improve their developer productivity and experience (`https://backstage.spotify.com/blog/how-spotify-measures-backstage-roi/`). An IDP was also captured in the `PTE.Platform.DeveloperCentric` item under the *Iteration 1* transition state. Prakash did some analysis and identified Backstage as the IDP of choice for the Resurgence platform after consulting with the core team. He updated the Resurgence platform capabilities mapping diagram with these details and added it to the wiki page. The following diagram shows the updated mapping for the IDP:

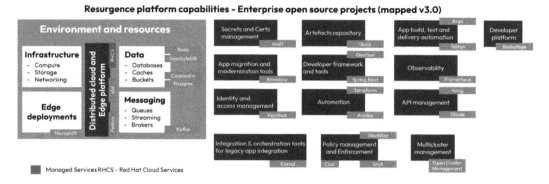

Figure 7.10 – Resurgence platform mapping with IDP mapping

Prakash notified the team on the channel about this addition. Later in the day, Ian looked at the updated mapping diagram and verified that all his questions were addressed. He pinged the team on the core team channel that he had reviewed the changes and the architecture seemed to address all his questions. He asked the core team to proceed with the review process.

Prakash initiated the review process with the Platform and Tenant Experience stream team over a 90-minute meeting the following week. In the meantime, Prakash and Susan worked on building the enterprise open source products mapping architecture that maps the capabilities to enterprise open source products, as documented in the `ResurgencePlatformTechnologiesMapping-Finalized` document. They started by grouping related products together since some products were mapped to multiple capabilities. The following screenshot shows the products-to-capabilities mapping:

Red Hat product	Capabilities provided
Red Hat OpenShift Platform Plus (RH OPP)	Container platform, OpenShift virtualization, image registry (Artefacts repo), multicluster management, policy management and enforcement, container data management
Red Hat Application Foundations (RHAF)	Java application frameworks, API connectivity, in-memory distributed datastore(cache), MTA, integration, single sign-on, Java application servers
Red Hat Device Edge (RHDE)	Edge optimized OS, lightweight container orchestrator

Figure 7.11 – Red Hat products mapped to capabilities

With this information, Prakash and Susan proceeded to build the enterprise open source products mapping architecture diagram, as shown here:

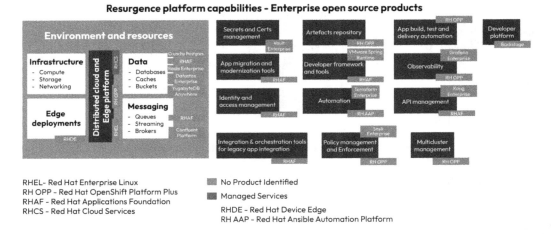

Figure 7.12 – Resurgence platform enterprise open source product mapping

The stream team reviewed the architecture diagram during this meeting. They realized that there was no reliable enterprise open source product available yet for Backstage; all the products they evaluated, including Red Hat Developer Hub, were either still not ready or not yet proven in the market. They decided to leave the Backstage product mapping for the next iteration and captured the decision in the decision log. They agreed that the `ResurgencePlatformTechnologiesMap ping-Finalized` document addressed all items listed under the `PTE.Platform` dimension and approved it for implementation.

During this meeting, the stream team brought up the availability of validated patterns, a set of deployment patterns including working examples maintained by Red Hat. Susan acknowledged her familiarity with the validated patterns and shared with the team that the Multicluster DevSecOps pattern (`https://hybrid-cloud-patterns.io/patterns/devsecops/#pattern-logical-architecture`) seemed to cover most of the platform capabilities that they had captured for the Resurgence platform and also leveraged the same set of enterprise open source products. The stream team asked Susan to work with the core team to learn more about this pattern and bring it up during the implementation architecture review workshop.

With the approval in place, Prakash updated the `PTE.Platform` sheet to reflect the implementation approval decision by changing the percentage complete to 40%, as per the **Definition of Done table** definition. The following screenshot shows the updated `PTE.Platform` sheet with items marked at 40% complete:

Dimension	Dimension Item	Item Description	Percentage Complete
PTE.Platform	UnifiedPlatform	A comprehensive and centralized platform that provides a consistent and seamless experience for customers, employees, and partners across all touchpoints and channels.	40%
PTE.Platform	Interoperability	Platform should be able to interact with other systems and applications in the technology ecosystem. It should use open standards and APIs to facilitate integration with other systems, making it easier to share data and functionality across the organization.	40%
PTE.Platform	Scalability	Platform should be designed to scale horizontally by adding more resources (such as servers or databases) to increase capacity as needed.	40%
PTE.Platform	Flexibility	Platform should be able to evolve and improve over time, incorporating new features and functionality as required.	40%
PTE.Platform	Security	Platform should incorporate robust authentication and access control mechanisms to protect sensitive data and prevent unauthorized access.	40%
PTE.Platform	Analytics	Platform should provide analytics enabling organization to track user behavior, measure the effectiveness of the digital initiatives, and make data-driven decisions.	40%
PTE.Platform	DeveloperCentric	Platforms should be developer-friendly, providing tools and resources that enable developers to build and deploy new workloads and services quickly and easily.	40%
PTE.Platform	Automation	Platform should utilize automation as much as possible to improve security, availability and resiliency.	40%
PTE.Platform	CloudAgnostic	Platform should provide a cloud agnostic architecture so that applications and workloads can be moved across different providers and between on-prem and cloud environments as required.	40%

Figure 7.13 – PTE.Platform with percentage complete updated to 40% for all items

The core team decided to go out for a drink later that day to celebrate the milestone. They invited Ian and Donna to join them. Donna said she could join them for the first hour but had some personal work to attend to after that. Ian said Fiona the CFO had set up an emergency meeting with him at the end of the day and he would swing by if that meeting ended in time. The core team insisted that Ian joined them after the meeting. The team left the office at 5 P.M. and walked to the Blues Bar right across from their Chicago office. Donna was already there waiting for them. While they were exchanging pleasantries, Donna's phone rang and it was Ian calling her. She told the team, "*I wonder what this is about? I do not typically get a call from Ian at this time of the day,*" and picked up the call. "*Hey Donna, sorry to bother you at this time. I know you are out with the team, but this is very important. If you are in a secluded place and if it is safe to talk, can you put me on the speaker so that I can talk to the team?*" asked Ian. Donna said, "*Give me a minute, Ian. The team is here with me. We will find a quiet place and call you from there,*" and disconnected the call.

It was not a busy time at the bar that evening. Prakash told the team that there was a dartboard behind the bar counter with a few tables and chairs and it did not get crowded till late in the night, and he took the team there. As expected, there was no one there, and the team was able to find the furthermost table and settled down quickly, wondering what this was all about. "*Hopefully the Resurgence program is not shelved,*" whispered Paul. After some giggles, Donna called Ian and put him on the speaker. Ian directly jumped into it: "*Listen, team, I had a not-so-good conversation with Fiona this evening. One of our tech vendors who has been with us for more than a decade has sent Fiona a $5.2M bill for out-of-compliance. Looks like our latest deployment model is not approved by the vendor, so we require more licenses than we purchased, even though our usage has gone down since we have moved most of the functionalities to our e-commerce site. Fiona said she is approaching the legal team to take a look at it, and no immediate action is required from this team. I just wanted to share this with you all,*" and paused.

The mood on the floor suddenly changed and an awkward silence crept in. Donna broke the silence and asked, *"Ian, does this impact the Resurgence program?"* Ian replied, *"No. However, I want to make sure we do not get into this situation with this program. Are we building a platform that is going to lock us in with a selected few vendors for the foreseeable future? I hope not,"* said Ian. The team looked at Donna, puzzled. Donna said, *"Ian, great question. The team is puzzled as to why I have not already communicated this to you. I think we do not have to worry about this situation happening in the future, or—should I say—the probability of this happening with the Resurgence platform is minimal. The reason why the team and I are confident about this is that the Resurgence platform is based on enterprise open source technologies. Even though there are certain tools that do have some proprietary features built in, as I explained in our last meeting, the core of the platform is built on open source, supported, and—in the case of cloud services—managed by Red Hat. The key benefit of this approach is that when we align with the open source ecosystem, we are working with an innovation model that is now the industry standard, with most tech vendors contributing to it, and not just a hotchpotch of technologies that we have to put engineering resources toward in order to make it work together. Hope this addresses your concern,"* said Donna. Ian thought about it and asked, *"Donna, do you think we should spend some time thinking about an alternate approach as well so that we can compare and contrast before selecting the best option for the Resurgence platform?"* Donna said, *"We can definitely explore that option if you want us to. However, based on our initial analysis as part of the platform capabilities mapping, there is no out-of-the-box alternative option for the Resurgence platform that covers all required capabilities. We can certainly build our own platform and leverage tools from various vendors and open source technologies; however, I think that it is a much bigger risk. So, to simplify things, we can consider either approach to have some risks. However, going with an enterprise-grade open source approach is a* **progressive** *model where we are committed to an innovation model that is accepted as the industry standard by most tech vendors and is progressing at a greater pace. On the other hand, building our own platform is a* **regressive** *model, with a handful of our engineers focusing primarily on creating the platform and reinventing the wheel. This approach can eventually break down as the components of the platform evolve independently and the tech debt and security issues start piling up. Hope this helps,"* closed Donna.

Ian took some time to respond and said, *"Donna and team, great job on this. I am quite comfortable with the explanation. Paul, can you make sure this is captured clearly in our decision log for future reference? Also, please send a note to the stream teams on this discussion so that they are aware of it."* Paul acknowledged and Ian ended the call, and as the team walked back to the bar table, they saw Ian entering the bar, typing away happily on his phone. He finished typing and told the team that he just sent a note to Ed and Fiona about the thorough work the team has done so far. And with that, the evening started off on a high note and with cheers and celebration!

Summary

In this chapter, we showed how enterprise open source technologies can be leveraged to build Bison Insurance's distributed cloud and edge platform based on the operating model previously defined in *Chapter 6*, *Your Distributed Technology Operating Model in Action*.

In the next chapter, we will summarize our learning in this book and share some parting thoughts on how you can leverage this book and the templates to create your distributed cloud and edge operating model. Next, please find some additional resources for your reference.

Further reading

Refer to the following links for more information about the topics that were covered in this chapter:

- Red Hat open source strategy: https://www.redhat.com/en/about/open-source
- Red Hat Device Edge: https://www.redhat.com/en/technologies/device-edge?extIdCarryOver=true&sc_cid=701f2000001Css5AAC
- Red Hat OpenShift Platform Plus: https://www.redhat.com/en/resources/openshift-platform-plus-datasheet
- Six application migration strategies: https://aws.amazon.com/blogs/enterprise-strategy/6-strategies-for-migrating-applications-to-the-cloud/
- Lightweight Kubernetes distributions for edge computing: https://programming-group.com/assets/pdf/papers/2023_Lightweight-Kubernetes-Distributions.pdf
- CNCF landscape: https://landscape.cncf.io/
- MoSCoW method: https://en.wikipedia.org/wiki/MoSCoW_method

8
Into the Beyond

In this book so far, you have learned about the fundamentals of an operating model, been given a technology landscape overview, and read about previous industry attempts to bring structure to operating models. We then leaned into why the future is distributed and the different classifications of edge computing environments, before we presented a step-by-step guide on how to purpose-build your own operating model. To solidify what we have learned, we provided an implementation example based on an anonymized case study to help you apply the learnings in your organization.

In this chapter, we will cover the following additional topics:

- Operating models for multiple independent operating units across the globe
- Antifragility
- Gap analysis
- Technical debt
- Undifferentiated heavy lifting
- How to measure progress
- Challenges associated with creating and applying your operating model
- Managing a multitude of options

By the end of this chapter, we will have rounded out and recapped the previous chapters for you to get started on your journey.

Operating model challenges

In this section, we will cover a few gotchas or things to be aware of when developing your operating model. If you've been in your organization for a while, then you probably are already aware of some and perhaps – even better – know how to address your challenges. If you are new or don't know how to tackle some of the challenges in your organization, then this section is for you.

We have already mentioned some of the challenges along the way and hence won't dive deep into them again. However, this section provides a brief recap.

There is no ready-made operating model that is fit for purpose, hence we wrote this book to help organizations and leaders to get the most value out of their distributed technology, skills, and process investment by creating their fit-for-purpose operating model.

The challenges of having knowledge gaps and undocumented vital information can have many root causes, such as organizational culture, acquisitions, strong growth periods, and so on. The best way to tackle this is by utilizing what we referred to in this book as *open practices*. This is what we have observed over many years as the best way to have people and teams focus on outcomes and collaborate and share information. All this is further detailed in Jim Whitehurst's book *The Open Organization*.

Becoming antifragile

The author Nassim Nicholas Taleb describes *antifragile* in his book with the same name as "*things that gain from disorder.*" While we all understand what fragile means, we don't really have a word that describes the opposite, so the author claims. The word *robust* might come to mind, but *robust* better describes things that don't change when exposed to disorder, such as uncertainty, variability, imperfect/incomplete knowledge, and error. Antifragility does take a different approach instead of just going further. *Antifragile* describes things that thrive, evolve, or get better when exposed to stressors or change, and open source software does exactly that.

We need to be more explicit, though, with what we mean when we say *open source software*. We are not talking about a public GitHub repo that is maintained by a handful of people. While impactful software might start that way, we really refer to *enterprise-grade* or *enterprise-ready* software. *Enterprise-grade* means supported by an established vendor and security response teams, and with training and support provided, while a large ecosystem of developers (paid and unpaid) evolves the features and functions of that software. Red Hat's Enterprise Linux and OpenShift are examples of pure open source-based products. There are also *open-core* approaches, which HashiCorp and MuleSoft take. They offer proprietary extensions upon an open source core and therefore, they deviate from the open source community's give-back commitment and weaken antifragility. On the other hand, pure open source software is inherently antifragile. Let me explain.

Open source software is often considered antifragile because it benefits from stressors and challenges, becoming stronger and more resilient as a result. The *antifragile concept*, as described by Nassim Taleb, refers to systems that not only withstand stressors but actually improve and grow stronger as a result of them.

One of the key aspects of open source software that makes it antifragile is the fact that it is transparent and can be scrutinized and improved by a large and diverse community of developers. This means that any weaknesses or vulnerabilities can be quickly identified and addressed, making the software better and less susceptible to future attacks.

Additionally, open source software benefits from a decentralized development model, where contributions can come from anywhere in the world. This means that if one contributor drops out or loses interest, there are others who can step in and continue the work. This makes the software more adaptable and less dependent on any one individual or organization.

Finally, the open source community is often highly collaborative, with developers sharing knowledge, tools, and resources freely. This creates a culture of continuous improvement and innovation, with new ideas and approaches being tested and refined in a dynamic and iterative process. This culture of collaboration and experimentation further strengthens the software and makes it more antifragile.

One example of an open source project that demonstrates antifragility is the Linux operating system. Linux is an open source operating system that was initially created by Linus Torvalds in 1991. Since then, it has been continuously developed and improved by a large community of developers worldwide.

Here are some specific ways in which Linux demonstrates antifragility:

- Decentralized development
- Iterative development
- Adaptability
- Community support

Overall, Linux demonstrates antifragility because it is able to adapt and improve in response to changing circumstances, market demands, and user needs. This is thanks to the decentralized, iterative, and adaptable nature of its development process, and the support of its large and dedicated community and vendors such as Red Hat or Suse.

That's why we believe funding the open source ecosystem is by far the best way to tackle the complexities of a distributed future. And no, that doesn't include hyperscalers that clone or fork open source projects and offer it as a service without contributing back or open sourcing their operational code.

So, while some are still talking about being robust (preventing failure from known and expected stressors) or moving to resilience (quick recovery from disruptions), the evolution of this is to become antifragile. Nassim Taleb's suggestion goes even further: organizations need to develop the ability to move from a **post-traumatic stress disorder** (**PTSD**) or preventing failure mindset to a **post-traumatic growth** (**PTG**) or opportunity-to-improve mindset. And for that to happen, we need blameless post-mortems, tracer pills, and an antifragile continuous learning mindset embracing things gone wrong as great learning opportunities.

You can achieve this by recognizing and celebrating learnings from critical errors and working them into your operating model iteratively as part of your feedback loop dimension.

We will now look into how open source, antifragile, undifferentiated heavy lifting, and tech debt are connected.

Understanding how open source connects antifragile, undifferentiated heavy lifting, and tech debt

It is important to recognize that undifferentiated heavy lifting and technical debt are different from antifragility but related.

Undifferentiated heavy lifting is an effort that the customer doesn't notice or care about. For example, whether your favorite bank builds its own server operating system or uses an established vendor's operating system makes little difference to the online banking user experience, if things go well. However, it exposes the organization and its customer to more cost and risk because the bank's core business isn't building operating systems but providing banking and financial services industry-related products and services. While I don't know any bank that builds its own operating systems, there are banks that are cobbling together fundamental services such as IAM, storage, logging, and monitoring with either their DIY or a subscribed naked Kubernetes service. Fundamentally, that's equivalent to building your own operating system. This is undifferentiated heavy lifting, costly and risky. And it builds up tech debt.

And worse, from a customer's perspective, instead of evolving banking products and services that the customer would benefit from, mortgage payments and bank fees paid by the customers are used to fund that undifferentiated heavy lifting.

We have already covered how open source software is inherently antifragile. With ready-made, Kubernetes-based, open source cloud platforms available, we can now make a case that open source limits undifferentiated heavy lifting and reduces tech debt if those platforms are based on community-driven open projects instead of forks. That is because those projects evolve in a meritocratic way, with the design trade-offs made in a best for all way. Most of the time, at least.

Technical debt refers to the costs that accrue over time when shortcuts or compromises are made in software development in order to meet short-term deadlines or achieve other goals. This can lead to code that is difficult to maintain, update, or scale in the future, and can result in higher costs and reduced efficiency over time. This can easily happen if you take random services and open source projects and cobble them together. It also happens when you reinvent the wheel by trying to build things in parallel with existing open source projects instead of contributing to those projects. I witnessed platform projects in many organizations ultimately being phased out when the key people left. On the other hand, technical debt can be necessary to make decisions and get things done, business value proven, or market opportunities seized.

Lastly, we now argue that by rebuilding features such as an enterprise cloud, edge, or Kubernetes platform, organizations are doing themselves a big disservice. Equivalent open source projects evolve just like any open source project evolves in an antifragile manner; companies that are not reusing enterprise-ready open source cloud and edge platforms are hence effectively financing the build-up of technical debt by cutting themselves off from community-driven innovation. That includes especially lego-ing your own platforms together.

"But those enterprise-ready platforms are opinionated!" I hear you say. Yes, and if you look at a snapshot of the **Cloud Native Computing Foundation** (**CNCF**) landscape in the following figure, it hopefully becomes clear why you want, or even need, an opinion. The question then is: which opinion do you trust most? Is it your organization's core business to stay abreast of those opinions and have an up-to-date opinion and the ability to influence and support (with engineering resourcing) the direction of those opinions?

If not, then it's probably best to listen to the opinions of organizations whose full-time job it is to do that.

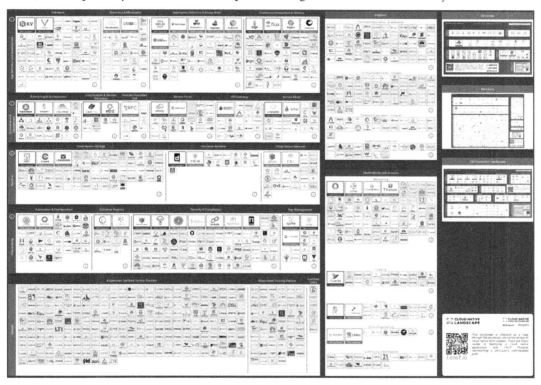

Figure 8.1 – CNCF landscape depicting the tools and technologies to choose from

But even that needs to be differentiated further:

- Does the opinionated stack come from a single vendor that wants to sell you services encouraging you to "keep building" (your lock-in, undifferentiated heavy lifting, and tech debt)?

- Are those opinions based on a chosen few within your organization who try to keep up with the latest developments and trends in the marketplace with a limited view of all the security and compliance requirements and the evolution thereof?

- Are those opinions coming from an active open source ecosystem, supported by vendors who develop and support open source projects, and provide security response teams because they understand what "enterprise compute" requirements are?

So basically, we are saying there is no such thing as "unopinionated" – it simply does not exist. You just need to be aware of where opinions come from and which source you trust most or are most aligned with your compute requirements.

And you can take this one step further: your organization can work with an open source vendor in order for your customizations and enhancement to flow into open source projects. That way, what would usually end up as tech debt can be incorporated into open source projects, supported as vendor products, and therefore, reduce your technical debt. That's another aspect of the power of the open source ecosystem.

In summary, the use of enterprise-ready open source software can limit technical debt, reduce undifferentiated heavy lifting, and help organizations to become antifragile.

Measuring organizational progress

We believe the best way to measure progress in a distributed technology operating model is via *flow metrics*. If you baseline your percentage of complete and accurate lead, wait, and process times, you will not only be able to optimize individual processes but also see whether completely revamped processes would actually help improve the overall outcome. For example, if your compliance process still imposes a wait time of 4 to 6 weeks to get your compute instance provisioned, does it matter that the actual provisioning of an EC2 instance only takes 5 minutes versus the previous 1 hour on your local virtual machine farm?

A word of caution: do not use metrics to compare teams. Problem spaces and technology stacks impose various degrees of complexity, hence comparing the flow between the mainframe team and the digital-natives team is unlikely to provide any insights. And don't forget to consider building slack into your flow as Martin Fowler suggests, as it shows various promising stats around team responsiveness, productivity, and collaboration, just to name a few.

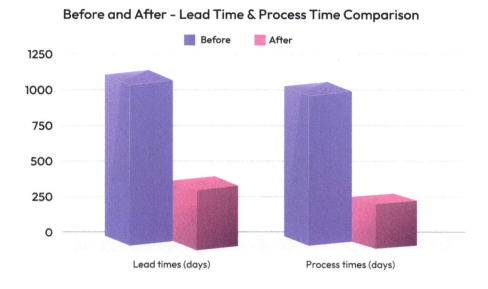

Figure 8.2 – Showing before versus after processing times (bank financial crime use case)

So far so good for the team metrics. For the overall operating model progress, we need something slightly different: some organizations like maturity models. I personally think there is a risk of creating a mindset of *arriving at a maturity level and we are done*. This sets the wrong expectation. However, if your organization is very familiar with maturity models, applies them well, and gets that the most-mature level is only the most-mature level for a short while before the next technology, process disruption, or regulatory compliance requirement comes along, then you could measure your technology operating model this way.

We covered capabilities in *Chapter 1*. Another way to measure progress is to define and measure capability increments along different capability axes such as people, processes, technology, or others. That way, it imposes less of a *reach the target state and we are done* mindset risk but focuses more on the multi-dimensionality of maturing your operating model as new capabilities are built and get expanded from DevOps to DevSecOps or define additional capability increments, as per the following example.

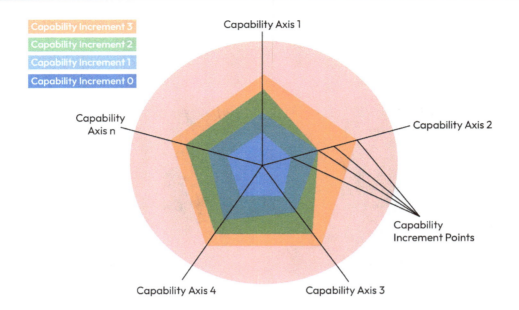

Figure 8.3 – An example capability model to measure progress along its axes via capability increments

Next, we are covering how to do your gap analysis in order to inform your roadmap.

Target operating model – gap analysis and roadmap

Once your dimensions are defined and scoped, you want to do a quick assessment of the current state as well as define the desired current target state of the dimension for progress reporting. My recommendation is to not dwell too long on the current state analysis. Your time is better spent defining suitable value-creating transitions and target states. In the *Measuring organizational progress* section, we introduced capabilities and capability increments. Hence, if you want to link your operating model with enterprise architecture or capability anchor maps, you don't have to introduce additional terms and language and potentially cause confusion. You can simply look at the in-scope items within your dimension, and document where you are at the moment, where you want to be, and what steps are necessary to get there. The following screenshot shows the target state for iterations 1, 2, and 3, based on the fact that your teams can now self-rate their progress and roll the item's progress percentage up. We repeat this process for all the operating model components (items, dimensions, and streams):

Notes	Iteration I- Transition Slate	Iterabon 2- - Transition State	Iteration 3- Transition State
Across Bison and Wisent, different platforms and technologies are used. They were adopted organically with little standardization work so far.	Create a unified platform architecture and migration plan to support all modern cloud native workloads and existing traditional workloads.	Build a unified cloud platform to address all business needs includng faster innovation and standardized operations.	Build and deploy modern cloud native workloads to the unified platform.

Figure 8.4 – Example transition state documentation

To ultimately get a picture of where the operating model is in terms of completeness, we roll up the individual line items of an operating model dimension into the dimension, and the dimensions into their respective streams. The different streams' cumulative progress depicts the overall progress and completion percentage of your operating model, as shown in the following screenshot:

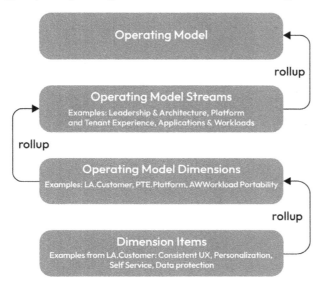

Figure 8.5 – Roll up for progress reporting of the different operating model components

To help teams judge and self-rate their progress, you can introduce a progress bar that teams can consult. It can be something simple as in the following screenshot:

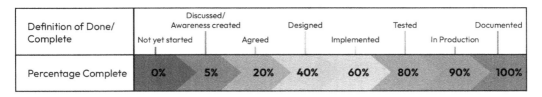

Figure 8.6 – Definition of Done per operating model item

However, the preceding figure is just a depiction of a single line item's iteration. If you are working on the `LA.Customer.ConsistentUX` line item with 5 transition states, then achieving 100% is equivalent to 100% divided by 5 transition states, which equals 20%. Subsequently, as a new target or new transition states are introduced for a line item (through your feedback loops, for example, or new business requirements or regulatory changes), the percentage complete will be lowered each time new target states are incorporated into your operating model.

The following is an operating model dimension progress snapshot across all items of the Customer Dimension:

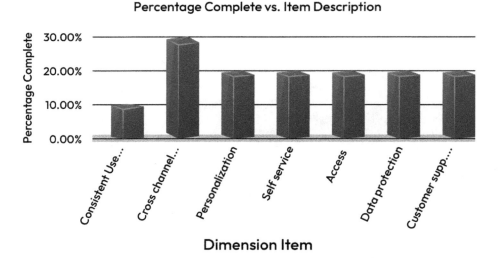

Figure 8.7 – Completeness shown for each line item (blue bars) per operating model dimension

This then rolls up into the streams' progress report tab, as shown:

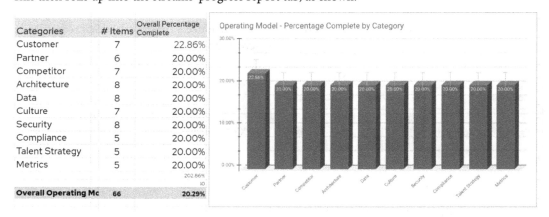

Categories	# Items	Overall Percentage Complete
Customer	7	22.86%
Partner	6	20.00%
Competitor	7	20.00%
Architecture	8	20.00%
Data	8	20.00%
Culture	7	20.00%
Security	8	20.00%
Compliance	5	20.00%
Talent Strategy	5	20.00%
Metrics	5	20.00%
		202.86%
		10
Overall Operating Mc	**66**	**20.29%**

Figure 8.8 – Operating model stream completion and progress reporting dashboard

And that in turn rolls up into the overall operating model completeness reporting dashboard, as shown:

Stream ID	Categories	Business Relevance & Purpose		# Topics	Overall Percentage Complete
1	Leadership & Architecture Stream	The Leadership & Architecture stream involves establishing clear goals, defining roles and responsibilities, and providing direction and guidance to the team. This includes setting a vision for the business, creating a strategy to achieve it, and managing and motivating the employees to execute that strategy effectively. Effective leadership can help to build a strong and resilient organization that can adapt to changing market conditions and emerging opportunities. It also involves designing and implementing the technical infrastructure that supports the digital business. This includes creating a technology roadmap that outlines the systems, tools, and platforms required to meet business objectives, and ensuring that these systems are scalable, secure, and reliable.		66	20.00%
2	Platform & Tenant Experience Stream	The Platform & Tenant experience stream involves developing and maintaining the technical infrastructure that supports the business, such as your website, mobile app, or other digital platforms and core applications. This includes selecting the appropriate technology stack, managing the software development process, and ensuring that your platforms are scalable, reliable, and secure. An effective platform can help you deliver a seamless user experience, enhance customer engagement, and drive business growth. It also involves creating a positive experience for customers who use your digital platforms or services. This includes designing user interfaces that are intuitive and easy to navigate, providing responsive customer support, and offering personalized experiences based on customer data and preferences. An effective tenant experience can help you build customer loyalty, increase retention rates, and drive revenue growth.		72	20.00%
3	Applications & Workloads	The Applications & Workloads stream involves developing, purchasing and maintaining the software applications that power the business. This includes developing custom applications, buying third party applications from independent software vendors, integrating third-party applications, and ensuring that your applications are scalable, reliable, and secure. Effective applications can help to streamline the operations, automate processes, and enhance the customer experience. It also involves optimizing the infrastructure that supports your applications and workloads. This includes selecting the appropriate cloud or on-premises infrastructure, adding edge infrastructure as needed, managing capacity and performance, and ensuring that your infrastructure is secure and compliant. Effective workloads can help you reduce costs, improve efficiency, and enhance the user experience for your customers.		58	20.00%
					60.00%
					3
	Overall Operating Model Completeness			**196**	**20.00%**

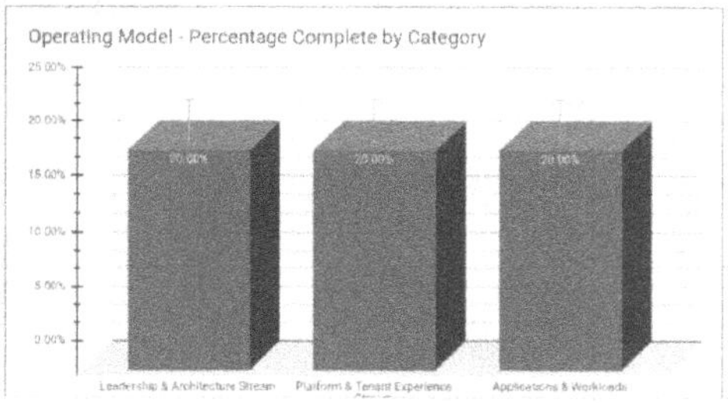

Operating Model - Percentage Complete by Category

Figure 8.9 – Operating model cumulative completion and progress dashboard

The template provided in our Git repository does the rollup automatically for you. You can examine the column formulas if you want to extend or understand the behavior in more detail.

We hope this section has provided some clarity with regards to how you can track and report your operating model progress. If you can utilize online spreadsheets serving office suites such as MS Teams, Excel, or Google Sheets, then you don't have to write lengthy progress update reports but simply share a link, which is what we recommend.

How much architecture do we need?

Just enough architecture is a principle in software development that emphasizes the importance of designing and implementing only the necessary architecture for a given project, without adding unnecessary complexity or overhead.

The idea behind *just enough architecture* is to strike a balance between having a well-designed and maintainable system, and avoiding over-engineering or wasting resources on features that may not be needed. This principle encourages people to focus on the most important aspects of the system, such as functionality, usability, and scalability, while avoiding excessive design or documentation that can slow down development and impede progress.

That said, organically-grown architectures with no governance or design principles are just as bad. The balance we suggest in regard to our operating model is that the architecture should consider all in-scope items in the current most evolved target iteration.

There are different ways to define architectural viewpoints from angles including business, information systems, and technology, such as **The Open Group Architecture Framework (TOGAF)**. Or if you go the Zachman route, then the architectural layers are called contextual, conceptual, logical, and physical. Because architecture's job is documenting the trade-offs you have chosen to solve a business problem, I prefer the latter; helping organizations set the context (scope) and then building out from conceptual to physical helps to refine and detail the thinking along the way:

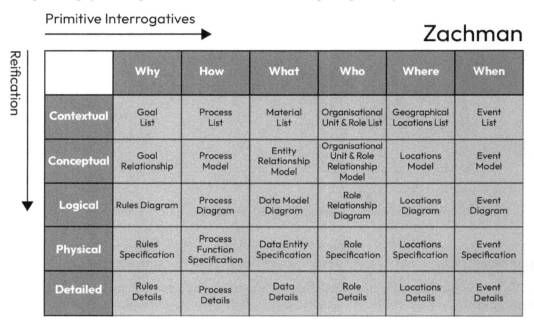

Figure 8.10 – Zachman architecture layers

Zachman also helps align teams and SMEs better to a business problem within a business domain (such as passenger vehicles compared to electric utility vehicles, for example). When I had to create my enterprise architecture team in a previous job, it would have been easy to follow the TOGAF path and dissect several business domains along business, applications and data, and infrastructure. But that would have meant that none of my architects would have been a suitable advisor to the LOB executives because of the limited understanding of how things fit together within a business domain. Hence, I think a domain-aligned structure where architects look after Order-to-Cash, Mine-to-Ship, or Call-to-Service domains, for example, yields much better business support.

The ivory tower mindset

"Mirror, mirror on the wall, who has the best operating model of them all?" As the fairy tale goes, it'll be your operating model for sure if you sit in your ivory tower in front of your own mirror all day. But if it's the best only because you developed it on your own or with your best friends and didn't share it for feedback, then it is actually a waste of time and effort. Firstly, it will most likely not be implemented, and secondly, adoption will not happen because your stakeholders haven't been involved:

Figure 8.11 – An ivory tower cartoon to lighten your mood

Sadly, I did work in an organization where big enterprise architectures were developed without involving stakeholders much. The longer that work went on, the more frightening it became to actually show those beautiful slides to anyone as the fear of rejection grew. Psychologically, this is probably because the amount of rework gets bigger the more you develop your thinking before getting feedback. When we finally shared our new (and, for the record, better) architectural approaches with project teams, you can imagine that the new approach developed in our ivory tower didn't find many friends. Even though, months later, the new architecture was endorsed at the executive level, the adoption still wasn't happening. Hence, we can't emphasize enough the importance of taking an open approach by using the open practices we have shared with you in this book. Utilizing open practices with your stakeholders is one vital key to a successful technology operating model rollout.

Understanding your organization's life cycles

The timespan of enterprise life cycles can vary depending on several factors, including industry, market conditions, technology advancements, and your enterprise's goals and strategies.

We covered strategies and tactics in relation to strategic goals and tactical objectives in *Chapter 1*. Strategies and tactics are the drivers behind the business model changes and new products being created or enhanced.

As you've learned throughout this book, your operating model is impacted by many different forces. However, from an enterprise life cycle perspective, products and services, business capabilities, and new, changed, or additional business models are the largest influencers creating pressure to evolve your operating model to support a new business strategy. The following figure shows how different enterprise life cycles have different timelines:

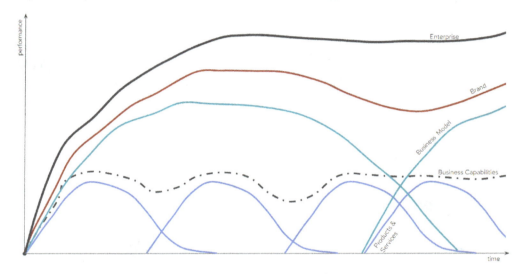

Figure 8.12 – A rough outline of timespans of enterprise life cycles and their correlation

We can generally compare the timespan of different enterprise life cycles as follows:

- **Business Model Life Cycle**: The timespan of the business model life cycle can be relatively long, as a business model can remain relevant and effective for many years. However, as market conditions change and new technologies emerge, a business may need to adapt its business model to remain competitive. This may involve significant changes to the enterprise's operations, marketing strategies, and product or service offerings, all of which can impact your technology operating model.

- **Business Capabilities Life Cycle**: The timespan of the business capabilities life cycle can also be relatively long, as developing core competencies and capabilities takes time and resources. However, as technology and market conditions change, businesses may need to adapt their capabilities to remain competitive. This may involve developing new skills or expertise, investing in new technologies, improving processes, or partnering with other businesses to fill gaps in their capabilities.

- **Products and Services Life Cycle**: The timespan of the products and services life cycle can be shorter than the other life cycles, as consumer preferences and market demand can change quickly. To remain competitive, businesses continuously innovate and update their product and service offerings to meet changing customer needs and preferences. New products or services, updating existing ones, or discontinuing products or services that are no longer in demand – your operating model needs to support that.

In summary, the timespan of enterprise life cycles varies. As the sponsor or owner of a technology operating model, your job is to ensure you are aware of coming changes and the triggers for those changes and to be in communication with your stakeholders who drive those changes. This is sometimes referred to as **sensing capability** and is vital for your technology operating model to remain fit for purpose.

Prioritization and making the right choice

When developing your operating model, there are a lot of choices and decisions to make. From stakeholder selection and grouping to operating model stream selection, the selection of stream dimensions, determining the scope, priority, and what technology and methodologies to favor only to welcome more input and requiring more decisions through working your feedback loops in – there is literally a never-ending set of decisions and choices to make. Often, we don't have all the necessary information or can't predict the future to know what will produce the best outcomes and argue the *priority slider* positions correctly. However, sometimes it helps to compare different options graphically and overlay them with all the principles, as it can be overwhelming remembering all the different aspects when casting your vote. For that, we wanted to show you a quick and useful colorful way to compare different options. The following figure shows green as supporting, black as not supporting, and gray as partly supporting the requirements on the left-hand side:

ID	Design Principles	OpenShift	Azure AKS	AWS EKS	Rancher
1	Reuse skills on-premises, in the public cloud and at the edge	green	black	black	gray
2	Establish a consistent security posture across all environments	green	black	black	gray
3	Limit undifferentiated heavy lifting across environments	green	black	black	gray
4	Ability to offload customization into open source projects	green	black	black	gray
5	Support near and far edge workloads	green	black	black	green
6	Natively integrated as public cloud services	green	green	green	black
7	Compliance controls out of the box	black	green	green	black
8	Align with current cloud skills	black	green	green	black
9	Support hybrid and multi cloud workload deployment requirements	green	black	black	gray
10	Other principles...				

Figure 8.13 – A quick graphical comparison tool to support decision-making in complex scenarios

Let's now move on to the more involved and sometimes divisive topic of lock-in.

Managing lock-in

Our good old friend the lock-in is back, with its most common first name, *vendor*. I've read books on cloud strategy where the author is trying to confuse readers with all types of different lock-ins, such as vendor, product, version, architecture, platform, skills, legal, and even mental lock-ins. I think, in reality, it's simpler. The ultimate litmus test of lock-in is this: If I stop paying, can I still use my systems and access my data? For example, if I pay an open source vendor for subscriptions or buy a perpetual license from a proprietary software vendor to use their products and I stop paying, then the product still runs and I can still access my data. I might no longer be able to call support or request a bug fix or patch, but my system is still running. If, on the other hand, I stop paying a SaaS provider or hyperscaler, then access to the system or infrastructure and data is cut off. Does the version still matter then? Or the architecture? No. This is ultimately the lock-in we need to be aware of, especially if that's either a dedicated dimension or a line item in one or more operating model dimensions as it creates dependencies into other dimensions and influences the architecture and technology choices you have.

On the other hand, this needs to be balanced with the fact that additional software subscriptions come with a price tag and that the implementation of additional software layers can lead to opportunity cost compared to using ready-made, fit-for-purpose, enterprise-grade hyperscaler proprietary services.

A related aspect is that cloud and edge platforms come with an ecosystem that can be hard to extend if certain offerings don't exist on your platform of choice. Hence, we recommend not just checking the technology but also how quickly the ecosystem grows and what type of third-party offerings are available.

From an operating model perspective, you can set guiding principles such as using the cloud as a resource not as a target location, using open source projects to abstract proprietary interfaces away, using only open source projects with a backing vendor, or buying only perpetual licenses, to name a few.

Multiple authorized operating model creators

So far, we have worked on the premise that organizations have established a single uncontested authority for the technology operating model design and execution. The reason for the existence of multi-cloud estates, however, can be based on the fact that different or specialized business units, geographies, or separate technology domains simply use their own preferred approach. Those specialized business units could be focused on **operational technology** (**OT**) or data science, could be business units with lots of autonomy such as home loans or payments, be business units acquired through acquisitions, or simply be in different geographies such as LATAM, EMEA, or APAC with separate reporting lines into HQ. Deliberately different infrastructure provider choices are made possible by PAYG procurement and credit card-based OpEx. However, it does create a challenge if you want to get a handle on your compliance and security posture, your data proliferation, storage cost, or cost in general.

If you work in an environment where you are tasked to tame an organically sprawling, multi-jurisdictional cloud or edge operating model, then the following figure shows a way to help analyze your operating model. You can categorize different aspects and get a handle on your as-is state in order to move into your desired target operating model.

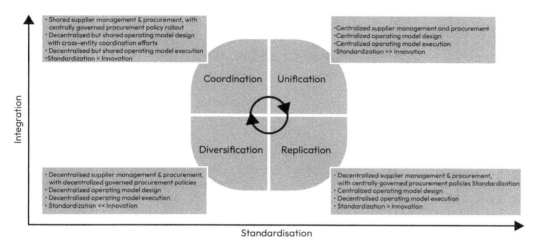

Figure 8.14 – Comparison of operating model design and execution modes

The preceding figure shows different operating model modes you can either drive toward or capture to tame.

Understanding Edge

If your background is in the IT domain, then it is possible that, at first, you find yourself struggling to explore the edge domain in your organization. That is mainly because the rules and priorities are different. For example (and a true story), in an organization, the IT department did a security scan of the OT network. Because the manufacturing equipment was set to use multicast to reduce network congestion, the scan alerted of a **distributed denial of service** (DDoS) attack. If the IT department would have been in charge, then the manufacturing plant would have come to a standstill with multiple millions of dollars in damages. That's one example of why Edge can require getting accustomed to different rules.

Another reason is that the language used can be different, plus the word *edge* might not even be used, even though the use case discussed is pure edge from an IT perspective. In a defense context, you might hear terms such as **Unmanned Aerial Vehicle** (UAV), task element HQ, or **Joint Task Force Headquarters** (JTF HQ). In a manufacturing environment, you might hear terms such as **MES**, **SCADA**, or **HMI** (referring to **manufacturing execution system**, **supervisory control and data acquisition**, and **human machine interface**) or be exposed to assumed knowledge such as understanding the difference between discrete and process manufacturing. In telco, you might hear terms such as **RAN**, **MEC**, or **BBU** (referring to **radio access network**, **multi-access edge compute**, and **baseband unit**).

That said, despite the difference, the telco vertical has started to consolidate IT and carrier-grade networking technology and processes and created an architecture called **horizontal telco cloud** to lay the foundation but also to use the same tooling such as automation and containers. The same consolidation efforts can be witnessed in manufacturing and other verticals.

For your operating model, that means that if specialized areas are in-scope, then you need to have the right SMEs, stakeholders, and change agents on board to incorporate the required knowledge about technologies and processes in use.

Sustainability

Measuring and offsetting carbon emissions is becoming an increasingly important topic for organizations. Many companies have carbon targets as part of their **Key Performance Indicators** (**KPIs**) to meet carbon-based commitments. Corporations are providing **Environmental, Social, and Governance** (**ESG**) statements, on top of financial statements, to investors and societies.

Even though we have not called out sustainability as one of the operating model dimensions in this book, it definitely is a great candidate with the increased focus on this topic from chip manufacturers to SaaS providers.

As energy efficiency gains have historically spurred more demand in information, communications, and operational services, which have exceeded the original efficiency gain-related savings, let's have a look at sustainability and approaches to help achieve carbon footprint reduction goals.

Some of the interesting questions in relation to sustainability in the ICT industry are the following:

- Are energy efficiency improvements in ICT continuing? (Undecided.)

- Is energy efficiency in ICT reducing carbon footprint? (Yes, but demand has increased, so No.)

- Are ICT emissions likely to stabilize due to saturation? (The number of devices people own might be, but with advances in edge computing and machine-to-machine communication, this might not be the case.)

- Is data traffic independent of ICT emissions? (No.)

- Is ICT enabling carbon savings in other industries? (It will depend on the extent to which ICT substitutes more carbon-intensive activities and whether ICT-enabled efficiency improvements in other industries might lead to a greater demand, which then offsets any efficiency gains.)

- Will renewable energy decarbonize ICT? (While the ICT sector is leading the way in the shift to renewable energy consumption, renewable energy itself has a significant carbon footprint across its supply chain.)

Microsoft published a report in which it illustrates the relationship between different scopes within the value chain:

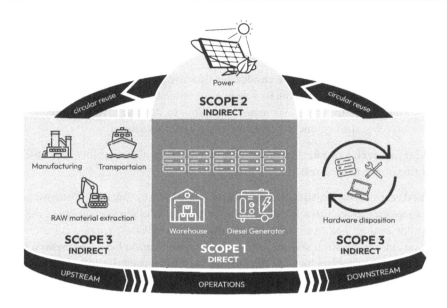

Figure 8.15 – Greenhouse gas emission scopes

The scopes are as follows:

- Emissions that directly result from business activities (scope 1)

- Emissions that result indirectly from producing energy (scope 2)

- Emissions that indirectly result from all other business activities inclusive of raw material extraction, manufacturing, and even recycling (scope 3)

The scope 3 emissions are usually much larger than scope 1 and 2 emissions. As a reference in 2020, the entire Microsoft value chain emitted 16 million metric tons of carbon. Scope 3 emissions can be about 40% thereof.

So, what can we do about it? First of all, if your organization is serious about climate action, then a sustainability dimension should be part of your distributed technology operating model. It is probably best placed in the leadership and architecture stream. From there, you can then ask for your supplier's actual carbon emission stats and their plans and actions to reduce the footprint of the products and services they sell. You can lengthen hardware refresh cycles from employee laptops to your data center and communications infrastructure hardware and you can select technology that allows you to reuse your existing infrastructure for cloud-native workloads across your extended hardware life cycle from on-premises, system Z, Power, ARM, and x86. You can look to increase workload density, reduce CPU frequency and idle cores, or move to greener data centers and clouds. Be aware that public cloud regions can have vastly different carbon footprints based on the energy generation means.

Summarizing our journey thus far

We have covered quite a bit of ground in this book.

Our path started off with looking at the challenges that the journey to the cloud has thrown at organizations and why – with edge making its way in every industry – we can't continue the same way we started off our journey to the cloud. We urged leadership to stay away from confusing marketing slogans such as *cloud-first strategy* and explained why we have this stance.

We demystified terms such as strategy, tactics, mission, vision, goals, capability, and objectives and explained how they relate to each other.

We then revisited research on organizational cultures completed by Dr. Ron Westrum and how to create performance-oriented cultures. We did this because it is the reason why we chose the workshop practices we outline in this book. And to be precise, this in turn is to help create a culture with a high level of collaboration, information sharing, and open communication because research has shown that those features are the enablers for high-performing teams and organizations. These are the same traits the open source movement displays.

After that, we took the operating model topic forward. We went through a rich set of examples from different consulting companies to look at their approaches and the way they define key components of an operating model: *the operating model dimensions*. The examples hopefully helped you build confidence in the fact that whichever dimensions you choose as the best-fit dimensions for your organization are automatically the most likely ones to be successful. In order to help you get a handle on the vast number of dimensions, we grouped them into streams. This will help you to assemble and group your stakeholders.

We then expanded the operating model topic into the cloud and edge realm and went into edge computing in more depth.

We defined key terms that we needed for the book but also stressed the fact that within a buzzword-rich industry, organizations are well advised to define a shared vocabulary across teams. Key terms included *engineering* and *operations*, distinguished between platform and product engineering and SRE, and we hopefully transformed some buzzwords along the way into tangible relevant topics for you. This is not just a mechanism to unlock organizational capabilities but also a foundational means to practice an inclusive high-performing culture because – with a common vocabulary – everyone is able to understand, discuss, and contribute with regard to what the organization is going through once semantics are attached to the terminology used.

The chapter got rounded out with thoughts on how to construct metrics and teaming, and how Conway's law affects your architecture.

Chapter 2 was dedicated to giving an overview of common enterprise technology landscape components. We distinguished between systems of innovation, differentiation, and systems of record before we examined the associated change cadence and challenges related to these cadences. We expanded the classification from an application point of view into the infrastructure realm and rounded the discussion out by looking into the difficulties around the adoption of a standard operating model. We even took two seemingly opposing approaches to reduce risk and improve stability and velocity, which are resisting change and embracing change.

In *Chapter 3*, we examined why the Gartner bimodal IT approach did not yield the results expected and extracted the learning out of it in order to apply it to our distributed technology operating model. We again looked at change cadence differences between mode 1 and mode 2 and looked at the tools, processes, and mindsets associated with it to find the root cause of the challenges and limitations.

Chapter 4 changed gears by starting to focus on the imminent distributed future – *imminent* because the future is already here, it's just not evenly distributed, so the saying goes. And we agree with that. Different industries are at different stages of technology and methodology adoption. We looked at the reasons why the future is distributed and revisited hybrid and multi-cloud definitions while double-clicking on specific business and technology reasons why the public or single cloud cannot be a target state for your operating model. We also spent time on different edge classifications, such as near, far, device or enterprise, operations, and provider, to get a better handle on different viewpoints in light of worthwhile use cases. We rounded the chapter out by looking at emerging trends and external factors such as compliance or mergers and acquisitions.

Chapter 5 is one of the core chapters of this book. It explains in detail the building blocks for a distributed operating model across the cloud and edge. We started off by showing the steps toward the desired outcome in the *Starting at the end* section via a flow chart and an operating model dashboard to track outcomes, **Work in Process** (**WIP**), and dependencies. We spent time on stakeholder management and showcased tools such as RA(S)CI and four-quadrant groupings to help you handle potential scale and complexity. We covered research on the perfect team size based on Dunbar's number, just before we presented workshop-leading practices to help lead teams along the forming, storming, norming, and performing life cycle. This is all in order to create a safe and open workshop environment to achieve the best outcomes for your operating model while working your dimensions with your stakeholder groups.

In addition, and in order to best work out the details of your operating model dimensions, we walked you through how to scope and group the dimensions into streams. We then walked through more than 30 dimensions for you to consider and choose from for your operating model, including the concerns those dimensions can cover as part of their scope. We also provided numerous suggestions for further reading if you want to dive deeper into any of the research underpinning our recommendations.

Chapter 6 is the other core chapter in relation to the development of your operating model. If you smell a rubbery smell while reading it, it is because that's where the rubber hits the road. We introduced an anonymized real-life use case and walked you through how this organization built its distributed technology operating model in a hybrid multi-cloud and edge context.

We did this by introducing the company, the market context, internal constraints, and mindset and then working with you step-by-step through the process, utilizing already introduced templates and new assets you can reuse for your operating model development process. We hope you find the documentation of the operating model dimensions, the iterations, and the definition of the (interim) target states with an inbuilt semi-automatic percentage-based gap analysis for you to measure your operating model completeness useful.

Chapter 7 walked through an operating model-based platform implementation example. It connected real world architecture, design, and implementation with the previously developed operating model. This chapter showed how requirements and principles from the operating model flow into technology selection and map to capabilities.

Chapter 8 wraps the book up. We introduced additional aspects that might impact or even evolve your thinking such as antifragility, geographically disparate (non-)autonomous operating models, different ways to measure progress, how tech debt, undifferentiated heavy lifting, and open source are connected, gap analysis, and roadmap development. On top of that, we discussed how much architecture we really need, sharing some experiences from within an ivory tower, understanding organizational life cycles, and comparing them with each other in the context of an enterprise life span. We revisited prioritization and decision-making and introduced a graphical way to make the best possible decision with the often limited information you have available. Finally, we close the book with this summary and will point to further reading underpinning our position outlined in this book.

What's next?

Firstly, you need to decide whether you need help or are confident to start this journey for your organization by yourself.

If you are ready to start on your own, then start socializing your idea with your colleagues, get executive sponsorship, and follow what's laid out in *Chapters 5, 6, and 7*. Kickstart your initiative with the assets made available in our GitHub repository (`https://github.com/PacktPublishing/Technology-Operating-Models-for-Cloud-and-Edge`).

If you need help for the short or medium term, then you can get help from consultants or your partners or contact the authors to help you find a suitable way to get started. Long-term ownership needs to sit within your organization, hence there's no need to seek long-term support from external entities. Note, if you get consultants and partners, make sure that those individuals have actually built and owned what they built beyond day 2 operations before. There is no point in getting help from people writing a very expensive 300-page document on your watch, which you are then left alone with to implement. Get people who've done it before with an extensive set of knowledge and experience that will flow into your operating model build-out.

Summary

And with that, we thank you for reading this far and wish you good luck on your path to defining and building your distributed future according to your needs! If you love the acceleration of new concepts and technologies, the increasing global connectedness, and the resulting shared learning as we do, then the conclusion can only be that it's an exciting time to be alive! Enjoy the enormous growth and learning potential, building high-performing teams, and leading with the culture aspects by example, and don't forget to share what you have learned by raising issues or pull requests on our GitHub repo.

Further reading

Refer to the following for more information about the topics covered in this chapter:

- Enterprise Architecture as Strategy: `https://hbsp.harvard.edu/product/8398-PDF-ENG`

- *The Open Organization*: `https://www.redhat.com/en/explore/the-open-organization-book`

- Martin Fowler on Slack: `https://martinfowler.com/bliki/Slack.html`

- The climate impact of ICT, Lancaster University: `https://arxiv.org/pdf/2102.02622.pdf`

- *Microsoft: A new approach for Scope 3 emissions transparency*: `https://download.microsoft.com/download/7/2/8/72830831-5d64-4f5c-9f51-e6e38ab1dd55/Microsoft_Scope_3_Emissions.pdf`

- Sustainability: `https://eprints.lancs.ac.uk/id/eprint/150985/1/TheClimateImpactOfICT_2020Report.pdf`

Acknowledgments

This book wouldn't have been possible without the help of others. Our employers were the very first to give us work, which helped us gain insights to write this book. We would also like to thank our families for giving us space and time to go through this long 6+ months journey.

But most importantly, our reviewers. We carefully selected our reviewers because of their expertise but also because of their personalities and the fact that our reviewers will all call out things that "don't seem right." All this is important to create a book of value for our readers, our industry, and the economy that thrives on business value creation. For that, we'd like to particularly thank the following:

- Katherine Squire, Senior Vice President, Engineering, Culture Amp

- Thenna Raj, Head of IT, Woolworths

- Guillaume Poulet-Mathis, Director of Product Engineering, Optus

- John Heaton, Chief Technology Officer, Alex Bank

- Niraj Naidu, APJ Head of Field Engineering, Datastax

- Sharad Gupta, Director of Sales Engineering, AMER Enterprise, UiPath

For us, it was important to secure the best technical reviewers with diverse technical and business backgrounds. This means that by combining the professional experiences of both authors and reviewers, you get access to knowledge from organizations such as Nasdaq, Zendesk, IBM, Red Hat, Draexlmaier Group, Webdefender, KraussMaffei, Wegmann, Information Builders, ANZ Bank, Fuji Xerox, Dell, VMWare, Pivotal, Oracle, Vision Point Systems, Celgene, BEA Systems, Philips Consumer Electronics, Culture Amp, nbnco, Australian Security Exchange, IRESS, Macquarie Group, Bankers Trust, Sydney Futures Exchange, Optus, Woolworths, KPMG, UXC, Alex Bank, Moneycatcha, Heritage Bank, Pivotal, Oracle, Datastax, and UiPath.

Index

Packtpub.com

Subscribe to our online digital library for full access to over 7,000 books and videos, as well as industry leading tools to help you plan your personal development and advance your career. For more information, please visit our website.

Why subscribe?

- Spend less time learning and more time coding with practical eBooks and Videos from over 4,000 industry professionals
- Improve your learning with Skill Plans built especially for you
- Get a free eBook or video every month
- Fully searchable for easy access to vital information
- Copy and paste, print, and bookmark content

Did you know that Packt offers eBook versions of every book published, with PDF and ePub files available? You can upgrade to the eBook version at packtpub.com and as a print book customer, you are entitled to a discount on the eBook copy. Get in touch with us at customercare@packtpub.com for more details.

At www.packtpub.com, you can also read a collection of free technical articles, sign up for a range of free newsletters, and receive exclusive discounts and offers on Packt books and eBooks.

Other Books You May Enjoy

If you enjoyed this book, you may be interested in these other books by Packt:

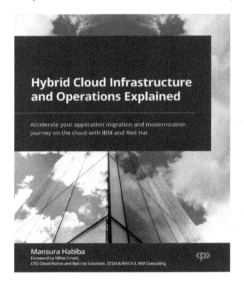

Hybrid Cloud Infrastructure and Operations Explained

Mansura Habiba

ISBN: 9781803248318

- Strategize application modernization, from the planning to the implementation phase
- Apply cloud-native development concepts, methods, and best practices
- Select the right strategy for cloud adoption and modernization
- Explore container platforms, storage, network, security, and operations
- Manage cloud operations using SREs, FinOps, and MLOps principles
- Design a modern data insight hub on the cloud

Azure for Developers - Second Edition

Kamil Mrzygłód

ISBN: 9781803240091

- Identify the Azure services that can help you get the results you need
- Implement PaaS components – Azure App Service, Azure SQL, Traffic Manager, CDN, Notification Hubs, and Azure Cognitive Search
- Work with serverless components
- Integrate applications with storage
- Put together messaging components (Event Hubs, Service Bus, and Azure Queue Storage)
- Use Application Insights to create complete monitoring solutions
- Secure solutions using Azure RBAC and manage identities
- Develop fast and scalable cloud applications

Packt is searching for authors like you

If you're interested in becoming an author for Packt, please visit authors.packtpub.com and apply today. We have worked with thousands of developers and tech professionals, just like you, to help them share their insight with the global tech community. You can make a general application, apply for a specific hot topic that we are recruiting an author for, or submit your own idea.

Share Your Thoughts

Now you've finished *Technology Operating Models for Cloud and Edge*, we'd love to hear your thoughts! Scan the QR code below to go straight to the Amazon review page for this book and share your feedback or leave a review on the site that you purchased it from.

https://packt.link/r/1837631395

Your review is important to us and the tech community and will help us make sure we're delivering excellent quality content.

Download a free PDF copy of this book

Thanks for purchasing this book!

Do you like to read on the go but are unable to carry your print books everywhere? Is your eBook purchase not compatible with the device of your choice?

Don't worry, now with every Packt book you get a DRM-free PDF version of that book at no cost.

Read anywhere, any place, on any device. Search, copy, and paste code from your favorite technical books directly into your application.

The perks don't stop there, you can get exclusive access to discounts, newsletters, and great free content in your inbox daily

Follow these simple steps to get the benefits:

1. Scan the QR code or visit the link below

https://packt.link/free-ebook/9781837631391

1. Submit your proof of purchase
2. That's it! We'll send your free PDF and other benefits to your email directly

www.ingramcontent.com/pod-product-compliance
Lightning Source LLC
Chambersburg PA
CBHW080523060326
40690CB00022B/5010